AF361682

Polynomials

Polynomials

Special Polynomials and Number-Theoretical Applications

Editor

Ákos Pintér

MDPI • Basel • Beijing • Wuhan • Barcelona • Belgrade • Manchester • Tokyo • Cluj • Tianjin

Editor
Ákos Pintér
Institute of Mathematics,
University of Debrecen
Hungary

Editorial Office
MDPI
St. Alban-Anlage 66
4052 Basel, Switzerland

This is a reprint of articles from the Special Issue published online in the open access journal *Symmetry* (ISSN 2073-8994) (available at: https://www.mdpi.com/journal/symmetry/special_issues/Special_Polynomials_Number_Theoretical_Applications).

For citation purposes, cite each article independently as indicated on the article page online and as indicated below:

LastName, A.A.; LastName, B.B.; LastName, C.C. Article Title. *Journal Name* **Year**, *Volume Number*, Page Range.

ISBN 978-3-0365-0818-4 (Hbk)
ISBN 978-3-0365-0819-1 (PDF)

© 2021 by the authors. Articles in this book are Open Access and distributed under the Creative Commons Attribution (CC BY) license, which allows users to download, copy and build upon published articles, as long as the author and publisher are properly credited, which ensures maximum dissemination and a wider impact of our publications.
The book as a whole is distributed by MDPI under the terms and conditions of the Creative Commons license CC BY-NC-ND.

Contents

About the Editor

Ákos Pintér is a Professor of Mathematics at the University of Debrecen. Over his career, he has obtained certain interesting results, among others, on the power values of power sums and applications of Baker and modular methods. His research concentrates on diophantine number theory, as well as general and special polynomials. His results have been published in more than 80 research papers. Pintér is a member of the Research Group of Number Theory in Debrecen, founded by Kálmán Győry, his supervisor. He is serving as the principal investigator of several scientific grants and the leader of research group Functions, Equations, Curves, sponsored by the Hungarian Academy of Sciences. In 2017, he received a shared Academic Prize for his mathematical achievements.

Preface to "Polynomials"

please add this part.

Ákos Pintér
Editor

Article

Two Variables Shivley's Matrix Polynomials

Fuli He [1], Ahmed Bakhet [1], M. Hidan [2] and M. Abdalla [2,3,*]

[1] School of Mathematics and Statistics, Central South University, Changsha 410083, China; hefuli999@163.com (F.H.); kauad_2006@csu.edu.cn (A.B.)

[2] Department of Mathematics, Faculty of Science for Girls, King Khalid University, Abha 61471, Saudi Arabia; mhedan@kku.edu.sa

[3] Mathematics Department, Faculty of Science, South Valley University, Qena 83523, Egypt

* Correspondence: moabdalla@kku.edu.sa

Received: 5 January 2019; Accepted: 21 January 2019; Published: 29 January 2019

Abstract: The principal object of this paper is to introduce two variable Shivley's matrix polynomials and derive their special properties. Generating matrix functions, matrix recurrence relations, summation formula and operational representations for these polynomials are deduced. Finally, Some special cases and consequences of our main results are also considered.

Keywords: Shivley's matrix polynomials; Generating matrix functions; Matrix recurrence relations; summation formula; Operational representations

MSC 2010: 33C05; 33C45; 33E20; 15A60

---

1. Introduction

Generalized Laguerre polynomials (GLP) are defined explicitly

$$L_n^a(x) = \sum_{r=0}^{n} \frac{(-1)^r \, (1+a)_n \, x^r}{r! \, (n-r)! \, (1+a)_r},$$
(1)

where a is a real -valued parameter, $(a)_r$ is the Pochhammer symbol

$$(a)_r = \begin{cases} a(a+1)\dots(a+(r-1)), & r \geq 1, \\ 1, & r = 0. \end{cases}$$

In confluent hypergeometric notation, we have

$$L_n^a(x) = \frac{(1+a)_n}{n!} \, {}_1F_1\big(-n; a+1; x\big).$$
(2)

These polynomials satisfy the second-order linear differential equation (see, for example, [1] p. 298)

$$x \, \mathbf{D}^2 L_n^a(x) + (1+a-x) \, \mathbf{D} L_n^a(x) + n L_n^a(x) = 0, \quad \mathbf{D} = \frac{d}{dx}.$$
(3)

The so-called Shively's pseudo-Laguerre polynomials $R_n(a, x)$ are defined by (see, [2])

$$R_n(a, x) = \frac{(a)_{2n}}{(a)_n \, n!} \, {}_1F_1\big(-n; a+n; x\big),$$
(4)

which are related to the proper simple Laguerre polynomial (see, [2])

$$L_n(x) = {}_1F_1(-n; 1; x),\tag{5}$$

$$R_n(a, x) = \frac{1}{(a-1)_n} \sum_{r=0}^{n} \frac{(a-1)_{n+r}\, L_{n-r}(x)}{r!}.\tag{6}$$

Shivley deduced the generating function for pseudo Laguerre polynomials of one variable as (see, [2])

$$e^{2t}\; {}_0F_1\left(-; \frac{a}{2} + \frac{1}{2}; t^2 - xt\right) = \sum_{n=0}^{\infty} \frac{R_n(a, x)\, t^n}{\left(\frac{a}{2} + \frac{1}{2}\right)_n}.\tag{7}$$

Now, owing to the significance of the earlier mentioned work related to Laguerre polynomials, we find record that many authors became interested to study the scalar cases of the classical sets of Laguerre polynomials into Laguerre matrix polynomials. Of those authors, we mention [3–7].

Recently, the matrix versions of the classical families orthogonal polynomials such as Jacobi, Hermite, Chebyshev, Legendre, Gegenbauer, Bessel and Humbert polynomials of one variables and some other polynomials were introduced by many authors for matrices in $\mathbb{C}^{N\times N}$ and various properties satisfied by them were given from the scalar case. Rather than giving an exhaustive list of references, we refer the reader to the article [8]. Theory of generalized and multivariable orthogonal matrix polynomials has provided new means of analysis to deal with the majority of problems in mathematical physics which find broad practical applications. In [9,10], Subuhi Khan and others introduced the 2-variable forms of Laguerre and modified Laguerre matrix polynomials and generalized Hermite matrix based polynomials of two variables and Lie algebraic techniques. Furthermore, several papers concerning the orthogonal matrix polynomials for two and multivariables have become more and more relevant, see for example [11–17].

The section-wise treatment is as follows. In Section 2, we deals with some basic facts, notations and results to that are needed in the work. In Section 3, we define Shivleys matrix polynomials of two variables and to study their properties. The generating matrix functions, matrix recurrence relations, summation formula and operational representations these new matrix polynomials are obtained. Some special cases of the established results are also underlined as corollaries. Finally, we give some concluding remarks in Section 4.

Throughout this paper, for $\mathbb{C}^N$ denote the N-dimensional complex vector space and $\mathbb{C}^{N\times N}$ denote all square matrices with N rows and N columns with entries are complex numbers, $\mathbf{Re}(z)$ and $\mathbf{Im}(z)$ denote the real and imaginary parts of a complex number z, respectively. For any matrix A in $\mathbb{C}^{N\times N}$, $\sigma(A)$ is the spectrum of A, the set of all eigenvalues of A, which will be denoted by $\|A\|$, is defined by

$$\|A\| = \sup_{x\neq 0} \frac{\|Ax\|_2}{\|x\|_2},$$

where for a vector y in $\mathbb{C}^N$, $\|y\|_2 = (y^H y)^{\frac{1}{2}}$ is Euclidean norm of y. I and $\mathbf{0}$ stand for the identity matrix and the null matrix in $\mathbb{C}^{N\times N}$, respectively.

2. Preliminaries

We shall adopt in this work a somewhat different notation and facts from that used throughout this work.

For $A \in \mathbb{C}^{N\times N}$, the matrix version of the symbol is

$$(A)^{(n)} = (A)(A - I) \cdots (A - (n-1)I), \quad n \geq 1,$$

and the Pochhammer symbol (the shifted factorial) is

$$(A)_n = A(A+I)\cdots(A+(n-1)I), \quad n \geq 1; \quad (A)_0 \equiv I.$$

Note that if $A = -jI$, where j is a positive integer, then $(A)_n = 0$ whenever $n > j$ (cf. [18]).

The reciprocal scalar Gamma function denoted by $\Gamma^{-1}(z) = \frac{1}{\Gamma(z)}$ is an entire function of the complex variable z. Thus, for any $A \in \mathbb{C}^{N \times N}$, Riesz-Dunford functional calculus [18–20] shows that $\Gamma^{-1}(A)$ is well defined and is, indeed, the inverse of $\Gamma(A)$. Furthermore, if

$$A + nI \quad \text{is invertible for all integer } n \geq 0, \tag{8}$$

then

$$(A)_n = \Gamma(A+nI)\Gamma^{-1}(A). \tag{9}$$

Form (9), it is easily to find that

$$(A)_{2n} = 2^{2n}(\tfrac{1}{2}(A+I))_n(\tfrac{1}{2}(A))_n,$$

and

$$(A)_{n+k} = (A)_n(A+nI)_k.$$

In 1731, Euler defined the derivative formula

$$D_x^{\nu} x^{\alpha} = \frac{\Gamma(\alpha+\nu)}{\Gamma(\alpha-\nu+1)} x^{\alpha-\nu}, \quad D_x \equiv \frac{d}{dx},$$

where α and ν are arbitrary complex numbers. By application of the matrix functional calculus to this definition, for any matrix $A \in \mathbb{C}^{N \times N}$, one gets (see [5,18])

$$\mathbf{D}_t^n[t^{A+mI}] = (A+I)_m[(A+I)_{m-n}]^{-1} t^{A+(m-n)I}, \quad \mathbf{D}_t = \frac{d}{dt}, \quad n = 0,1,2,3,\ldots$$

On other hand, if $\mathbf{D}_z = \frac{\partial}{\partial z}$, $\mathbf{D}_z = \frac{\partial}{\partial w}$ and $\mathbf{D}_v = \frac{\partial}{\partial v}$ the trinomial expansion for $(\mathbf{D}_z + \mathbf{D}_w + \mathbf{D}_v)^n$ is given by (see [21,22])

$$(\mathbf{D}_z + \mathbf{D}_w + \mathbf{D}_v)^n = \sum_{r=0}^{n}\sum_{s=0}^{n-r} \frac{(-1)^{r+s}(-n)_{r+s}}{r!\,s!} \mathbf{D}_z^{n-r-s}\,\mathbf{D}_w^r\,\mathbf{D}_v^s, \tag{10}$$

operating (10) on $F(z,w,v)$, we get

$$(\mathbf{D}_z + \mathbf{D}_w + \mathbf{D}_v)^n F(z,w,v) =$$

$$\sum_{r=0}^{n}\sum_{s=0}^{n-r} \frac{(-1)^{r+s}(-n)_{r+s}}{r!\,s!} \mathbf{D}_z^{n-r-s}\,\mathbf{D}_w^r\,\mathbf{D}_v^s F(z,w,v), \tag{11}$$

in particular, if $F(z,w,v) = f(z)g(w)h(v)$, then (11) gives

$$(\mathbf{D}_z + \mathbf{D}_w + \mathbf{D}_v)^n \left\{ f(z)g(w)h(v) \right\} =$$

$$\sum_{r=0}^{n}\sum_{s=0}^{n-r} \frac{(-1)^{r+s}(-n)_{r+s}}{r!\,s!} \mathbf{D}_z^{n-r-s} f(z)\,\mathbf{D}_w^r g(w)\,\mathbf{D}_v^s h(v). \tag{12}$$

Similarly,

$$(\mathbf{D}_z\mathbf{D}_w + \mathbf{D}_z\mathbf{D}_v + \mathbf{D}_w\mathbf{D}_v)^n \left\{ f(z)g(w)h(v) \right\} =$$

$$\sum_{r=0}^{n} \sum_{s=0}^{n-r} \frac{(-1)^{r+s}\,(-n)_{r+s}}{r!\,s!}\, \mathbf{D}_z^{n-s} f(z)\, \mathbf{D}_w^{n-r} g(w)\, \mathbf{D}_v^{r+s} h(v). \tag{13}$$

Moreover, if $A \in \mathbb{C}^{N \times N}$, and z is any complex number, then the matrix exponential e^{Az} is defined to be

$$e^{Az} = I + Az + \ldots + \frac{A^n}{n!}\, z^n + \ldots,$$

$$\frac{d^n}{dz^n}[e^{Az}] = A^n\, e^{Az} = e^{Az}\, A^n, \quad n = 0,1,2,3,\ldots.$$

Let A, B and C be matrices in $\mathbb{C}^{N \times N}$ and C satisfy condition (8), then the hypergeometric matrix function of 2-numerator and 1-denominator for $|z| < 1$ is defined by the matrix power series (see [20,23])

$${}_2F_1(A, B; C; z) = \sum_{n \geq 0} \frac{(A)_n (B)_n [(C)_n]^{-1}}{n!} z^n, \tag{14}$$

For an arbitrary matrix $A \in \mathbb{C}^{N \times N}$, satisfy condition (8) then the n-th Laguerre matrix polynomials $L_n^A(z)$ is defined by (see [8])

$$L_n^A(z) = \frac{(A+I)_n}{n!}\, {}_1F_1(-nI; A+I; z). \tag{15}$$

Therefore, the Shively's pseudo Laguerre matrix polynomials are reduced in the form

$$R_n^A(z) = \frac{(A)_{2n}\,[(A)_n]^{-1}}{n!}\, {}_1F_1(-nI; A+nI; z). \tag{16}$$

For matrices $A(k,n)$ and $B(k,n)$ are matrices in $\mathbb{C}^{N \times N}$ for $n \geq 0, k \geq 0$, the following relations are satisfied (see [24])

$$\sum_{n=0}^{\infty} \sum_{k=0}^{\infty} A(k,n) = \sum_{n=0}^{\infty} \sum_{k=0}^{[\frac{1}{2}n]} A(k, n-2k) \tag{17}$$

and

$$\sum_{n=0}^{\infty} \sum_{k=0}^{\infty} B(k,n) = \sum_{n=0}^{\infty} \sum_{k=0}^{n} B(k, n-k). \tag{18}$$

Similarly, we can write

$$\sum_{n=0}^{\infty} \sum_{k=0}^{[\frac{1}{2}n]} A(k,n) = \sum_{n=0}^{\infty} \sum_{k=0}^{\infty} A(k, n+2k), \tag{19}$$

and

$$\sum_{n=0}^{\infty} \sum_{k=0}^{n} B(k,n) = \sum_{n=0}^{\infty} \sum_{k=0}^{\infty} B(k, n+k). \tag{20}$$

3. Two Variables Shivley's Matrix Polynomials

In this section we define two variables Shively's matrix polynomials and several properties for these polynomials as given below:

Definition 1. *For an arbitrary matrix $A \in \mathbb{C}^{N \times N}$, with $A + mI$ invertible for every integer $m \geq 1$, then the m-th Shively's matrix polynomials of two variables $R_m^A(z, w)$ is defined by*

$$R_m^A(z, w) = \frac{(A + mI)_m}{m!} \sum_{n=0}^{m} \sum_{k=0}^{m-n} [(A + mI)_k]^{-1} (-mI)_{n+k} \frac{z^k \, w^n}{n! \, k!}. \tag{21}$$

Remark 1. *For simplicity, we consider only two complex variables Shively's matrix polynomials, though the results can easily be extended to several complex variables.*

3.1. Generating Functions and Recurrence Relations

Two more basic properties of two variables Shively's matrix polynomials are developed in this subsection. The generating matrix functions which is obtained from Theorem 1 and with the help of Definition 1. Also, some matrix recurrence relations for two variables Shively's matrix polynomials are given.

Theorem 1. *The generating matrix function of $R_m^A(z, w)$ is given by*

$$\sum_{m=0}^{\infty} R_m^A(z, w) \, [(\frac{A + I}{2})_m]^{-1} \, t^m = e^{2t} \, {}_0F_1\left(-; \frac{A + I}{2}; t^2(1 - w) - tz\right). \tag{22}$$

Proof. From Definition 1 in the left hand side of (22), we get

$$\sum_{m=0}^{\infty} R_m^A(z, w) \, [(\frac{A + I}{2})_m]^{-1} \, t^m =$$

$$\sum_{m=0}^{\infty} [(\frac{A + I}{2})_m]^{-1} \, t^m \, \frac{(A + mI)_m}{m!}$$

$$\times \sum_{n=0}^{m} \sum_{k=0}^{m-n} [(A + mI)_k]^{-1} (-mI)_{n+k} \frac{z^k \, w^n}{n! \, k!}$$

$$= \sum_{m=0}^{\infty} 2^{2m} \, (\frac{A}{2})_m \, (\frac{A + I}{2})_m \, [(A)_m]^{-1} \, [(\frac{A + I}{2})_m]^{-1} \, t^m$$

$$\times \sum_{n=0}^{m} \sum_{k=0}^{m-n} [(A + mI)_k]^{-1} (-mI)_{n+k} \frac{z^k \, w^n}{n! \, k!} \tag{23}$$

$$= \sum_{m=0}^{\infty} \sum_{n=0}^{\infty} \sum_{k=0}^{\infty} (\frac{A}{2})_n \, (\frac{A}{2} + nI)_{m+k} \, [(A)_{2n}]^{-1} \, [(A + 2n)_{m+k}]^{-1}$$

$$\times \frac{(-1)^{n+k} z^n w^k (4t)^{m+n+k}}{m! n! k!}$$

$$= e^{2t} \sum_{m=0}^{\infty} \sum_{n=0}^{\infty} [(\frac{A + I}{2})_{m+n}]^{-1} \frac{(t^2(1 - w))^m \, (-zt)^n}{m! \, n!}.$$

Further simplification yields

$$\sum_{m=0}^{\infty} R_m^A(z, w) \, [(\frac{A + I}{2})_m]^{-1} \, t^m = e^{2t} \sum_{m=0}^{\infty} [(\frac{A + I}{2})_m]^{-1} \frac{(t^2(1 - w) - zt)^m}{m!}$$

$$= e^{2t} \, {}_0F_1\left(-; \frac{A + I}{2}; t^2(1 - w) - tz\right). \tag{24}$$

This completes the proof of Theorem 1. □

Theorem 1 leads to the following corollaries:

Corollary 1. *The generating matrix function for the Shively's pseudo Laguerre matrix polynomials $R_m^A(z)$ is given by*

$$\sum_{m=0}^{\infty} R_m^A(z) \ [(\frac{A+I}{2})_m]^{-1} \ t^m \ = \ e^{2t} \ _0F_1\Big(-;\frac{A+I}{2};t^2 - tz\Big). \tag{25}$$

Proof. Follows by successive application of Theorem 1. □

Corollary 2. *From the generating matrix function (22), we can deduce that*

$$R_m^A(0,w) = \frac{1}{m} \ _2F_1\Big(-\frac{m}{2}I,\frac{-m+1}{2}I;I-(\frac{A}{2}+mI);w\Big). \tag{26}$$

Proof. By putting $z = 0$ in (22), we find

$$\sum_{m=0}^{\infty} R_m^A(0,w) \ [(\frac{A+I}{2})_m]^{-1} \ t^m$$

$$=e^{2t} \ _0F_1\Big(-;\frac{A+I}{2};t^2(1-w)\Big)$$

$$=e^{2t} \sum_{m=0}^{\infty} [(\frac{A+I}{2})_m]^{-1} \ t^{2m} \ \frac{(1-w)^m}{m!}$$

$$=e^{2t} \sum_{m=0}^{\infty} [(\frac{A+I}{2})_m]^{-1} \ \frac{t^{2m}}{m!} \sum_{k=0}^{m} \frac{(-mI)_k \ w^k}{k!}.$$

Further simplification yields

$$\sum_{m=0}^{\infty} R_m^A(0,w) \ [(\frac{A+I}{2})_m]^{-1} \ t^m$$
$$= \sum_{m=0}^{\infty} [(\frac{A+I}{2})_m]^{-1} \ t^m \ \frac{1}{m} \ _2F_1\Big(-\frac{m}{2}I,\frac{-m+1}{2}I;I-(\frac{A}{2}+mI);w\Big), \tag{27}$$

and the relation (27) evidently leads us to the required result. □

Among the infinitely many recurrence relations for two variables Shively's matrix polynomials, we list the following two as being the most useful or interesting ones.

$$AR_m^A(z,w) + z\mathbf{D}_z \ R_m^A(z,w) = (A+mI) \ R_m^{A-I}(z,w), \tag{28}$$

$$(A+mI) \ R_{m-1}^A(z,w) + (z\mathbf{D}_z + w\mathbf{D}_w) \ R_m^A(z,w) = m \ R_m^{A-I}(z,w), \quad \mathbf{D}_z \equiv \frac{\partial}{\partial z}, \quad \mathbf{D}_w \equiv \frac{\partial}{\partial w}. \tag{29}$$

It can be easily verifying these relations from the Definition 1.

3.2. Summation Formulas and Operational Representation

We are now in a position to obtain some series expansion formulae involving partial derivatives for the $R_m^A(z,w)$, these series expansion formulae are given by the following theorem:

Theorem 2. *Suppose that A is a matrix in $\mathbb{C}^{N \times N}$ satisfying (8) and $u \in \mathbb{C}$. Then the Shively's matrix polynomials of two variables has the following summation formulas*

$$\sum_{k=0}^{m} \frac{u^k}{k!} \, \mathbf{D}_z^k \, R_m^A(z,w) = R_m^A(z+u,w), \tag{30}$$

$$\sum_{k=0}^{m} \frac{u^k}{k!} \, \mathbf{D}_w^k \, R_m^A(z,w) = R_m^A(z,w+u), \tag{31}$$

$$\sum_{k=0}^{m} [(A+I)_{m-k}]^{-1} \frac{(-u)^k}{k!} \, \mathbf{D}_w^k \, R_m^{A-kI}(z,w)$$
$$= (1+u)^m \, [(A+I)_m]^{-1} \, R_m^A\Big(\frac{z}{1+u}, \frac{w}{1+u}\Big), \tag{32}$$

$$\sum_{k=0}^{m} \frac{(-u)^k((m+k)!)^2(-mI)_{-k}}{k!} \, \mathbf{D}_z^k \, \mathbf{D}_w^k \, R_{m+k}^{A-kI}(z,w)$$
$$= m! \, (1+u)^m \, R_m^A\Big(\frac{z}{1+u}, \frac{w}{1+u}\Big), \tag{33}$$

$$\sum_{k=0}^{n} \binom{n}{k} \, [(I+A-nI)_k]^{-1} \, z^k \, \mathbf{D}_z^k \, R_m^A(z,w)$$
$$= (A+I)_m \, [(I+A-nI)_m]^{-1} \, R_m^{A-nI}(z,w); \quad n \le m. \tag{34}$$

Proof. Taking the left hand side of (30) and substituting the value of $R_m^A(z,w)$ from (21), we get

$$\sum_{k=0}^{m} \frac{u^k(A+I)_m}{m!k!} \, \mathbf{D}_z^k \sum_{n=0}^{m} \sum_{r=0}^{m-n} [(A+mI)_r]^{-1} \, (-mI)_{n+r} \frac{z^r \, w^n}{n! \, r!}$$
$$= \frac{(A+I)_m}{m!} \sum_{n=0}^{m} \sum_{r=0}^{m-n} [(A+mI)_r]^{-1} \, (-mI)_{n+r} \frac{z^r \, w^n}{n! \, r!} \sum_{k=0}^{m} \frac{(-r)_k(\frac{-u}{z})^k}{k!}$$
$$= \frac{(A+I)_m}{m!} \sum_{n=0}^{m} \sum_{r=0}^{m-n} [(A+mI)_r]^{-1} \, (-mI)_{n+r} \frac{(z+u)^r \, w^n}{n! \, r!} = R_m^A(z+u,w). \tag{35}$$

This completes the proof of (30). Similarly, we can prove (31).

Taking the left hand side of (32), substituting the value of Shively's matrix polynomials of two variables from (21) and differentiating, we get

$$\sum_{k=0}^{m} [(A+I)_{m-k}]^{-1} \frac{(-u)^k}{k!} \, \mathbf{D}_w^k \, R_m^{A-kI}(z,w)$$
$$= \frac{1}{m!} \sum_{k=0}^{m} \frac{(-u)^k[(A+I)_{-k}]^{-1}}{k!} \sum_{n=0}^{m} \sum_{r=0}^{m-n} [(I+A-kI)_r]^{-1} \, (-mI)_{n+r} \frac{z^{r-k} \, w^n}{n! \, (r-k)!}. \tag{36}$$

Putting $r = \mu + k$ where μ is new parameter of summation and changing the order of summation so that the first summation becomes last,

$$\sum_{k=0}^{m} [(A+I)_{m-k}]^{-1} \frac{(-u)^k}{k!} \, \mathbf{D}_w^k \, R_m^{A-kI}(z,w)$$

$$= \frac{1}{m!} \sum_{n=0}^{m} \sum_{\mu=0}^{m-n} [(I+A-kI)_\mu]^{-1} (-mI)_{n+\mu} \frac{z^\nu \, w^n}{n! \, \mu!}$$

$$\times \sum_{k=0}^{m-n-\mu} \frac{(-m+n+\mu)(-u)^k}{k!}$$

$$= \frac{1}{m!} \sum_{n=0}^{m} \sum_{\mu=0}^{m-n} [(I+A-kI)_\mu]^{-1} (-mI)_{n+\mu} \frac{z^\nu \, w^n}{n! \, \mu!} (1+u)^{(m-n-\mu)}, \tag{37}$$

which in view of (21), gives us the right hand-side of assertion (32).

Also, from the left hand side of (33) and substituting the value of Shively's matrix polynomials of two variables from (21), we obtain

$$\sum_{k=0}^{m} \frac{(-u)^k (-mI)_{-k} (I+A-kI)_{m+k}}{k!} \, \mathbf{D}_z^k \, \mathbf{D}_w^k$$

$$\times \sum_{n=0}^{m+k} \sum_{r=0}^{m+k-n} [(I+A-kI)_r]^{-1} (-mI-kI)_{n+r} \frac{z^r \, w^n}{n! \, r!}. \tag{38}$$

Now, differentiating and substituting $p = r - k$, $q = n - k$, we have

$$(A+I)_m \sum_{k=0}^{m} \sum_{q=0}^{m-k} \sum_{p=0}^{m+k-q} [(A+I)_p]^{-1} \frac{(-u)^k (-mI)_{p+q+k}}{k!} \frac{z^p \, w^q}{p! \, q!}$$

$$=(A+I)_m \sum_{q=0}^{m} \sum_{p=0}^{m-q} [(A+I)_p]^{-1} \frac{(-mI)_{p+q} \, z^p \, w^q}{p! \, q!} (1+u)^{m-p-q}. \tag{39}$$

Again, using the expression (21), we arrive at the right-hand side of (33).
Consider the series

$$\sum_{k=0}^{n} \binom{n}{k} [(I+A-nI)_k]^{-1} z^k \, \mathbf{D}_z^k \, R_m^A(z,w)$$

$$= \frac{(A+I)_m}{m!} \sum_{k=0}^{n} \binom{n}{k} [(I+A-nI)_k]^{-1}$$

$$\times \sum_{n=0}^{m} \sum_{r=0}^{m-n} [(A+mI)_r]^{-1} (-mI)_{n+r} \frac{z^r \, w^n}{n! \, (r-k)!}$$

$$= \frac{(A+I)_m}{m!} \sum_{n=0}^{m} \sum_{r=0}^{m-n} [(A+mI)_r]^{-1} (-mI)_{n+r} \frac{z^r \, w^n}{n! \, (r!)^2} \, {}_2F_1\Big(-rI, -nI; I+A-nI; 1\Big).$$

In light of the relationship (see, [25])

$$F(-nI, B : C; 1) = \Gamma(C)\Gamma(C - B + nI)\Gamma^{-1}(C + nI)\Gamma^{-1}(C - B), \tag{40}$$

where $B, C \in \mathbb{C}^{N \times N}$, we obtain the required result in (34). $\square$

Remark 2. *Setting $n = 1$ in (34) we have the recurrence relations for the $R_m^A(z,w)$ in (28).*

Next, according to (13), we have the following operational representation for the $R_m^A(z, w)$:

$$
\left(\mathbf{D}_z\, \mathbf{D}_w + \mathbf{D}_z\, \mathbf{D}_v + \mathbf{D}_w\, \mathbf{D}_v \right)^m \left\{ z^{A+(2m-1)I}\, w^m\, e^{-v} \right\}
$$

$$
= \sum_{n=0}^{m} \sum_{k=0}^{m-n} \frac{(-1)^{n+k}\, (-mI)_{n+k}}{n!\, k!}
$$

$$
\times \mathbf{D}_z^{m-k}\left(z^{A+(2m-1)I} \right) \mathbf{D}_w^{m-n}\left(w^m \right) \mathbf{D}_v^{n+k}\left(e^{-v} \right)
$$

$$
= \frac{(m!)^2}{n!}\, z^{A+(m-1)I}\, e^{-v} \left\{ \frac{(A+mI)_m}{m!} \sum_{n=0}^{m} \sum_{k=0}^{m-n} \frac{[(A+kI)_k]^{-1}(-mI)_{n+k}\, z^k w^n}{n!k!} \right\}
$$

$$
= \frac{(m!)^2}{n!}\, z^{A+(m-1)I}\, e^{-v}\, R_m^A(z, w),
\tag{41}
$$

thus, we get

$$
\left(\mathbf{D}_z\, \mathbf{D}_w + \mathbf{D}_w\, \mathbf{D}_v + \mathbf{D}_v\, \mathbf{D}_z \right)^m \left\{ z^{A+(2m-1)I}\, w^m\, e^{-v} \right\}
$$

$$
= \frac{(m!)^2}{n!}\, z^{A+(m-1)I}\, e^{-v}\, R_m^A(z, w).
\tag{42}
$$

Summarizing, the following result has been obtained:

Theorem 3. *Let $R_m^A(z, w)$ be given in (21). The operational representation in (42) holds true.*

4. Concluding Remarks

This paper is to define a new matrix polynomial, say, Shivley's matrix polynomials of two complex variables and to study their properties. Some formulas related to an explicit representation, generating matrix functions, matrix recurrence relations, series expansion and operational representations are deduced. Also, some interested particular cases and consequences of our results have been discussed. Within such a context, new matrix polynomial structures emerge with wide possibilities of applications in physics and engineering. Therefore, the results of this work are variant, significant and so it is interesting and capable to develop its study in the future.

Author Contributions: All authors contributed equally.

Funding: This Work is Supported by National Natural Science Foundation of China (11601525), Natural Science Foundation of Hunan Province (2017JJ3406).

Acknowledgments: The third and fourth authors wish to acknowledge the approval and the support of this research study from the Deanship of Scientific Research in King Khalid University, Abha, Saudi Arabia under grant (R.G.P-215-39). The authors would like to thank the referee for the valuable comments, which helped to improve this manuscript.

Conflicts of Interest: The authors declare no conflict of interest.

References

1. Rainville, E.D. *Special Functions*; Chelsea Publishing Co.: Bronx, NY, USA; Macmillan: New York, NY, USA, 1997.
2. Shively, R. On Pseudo-Laguerre Polynomials. Ph. D. Thesis, University of Michigan, Ann Arbor, MI, USA, 1953.
3. Çekim, B.; Altin, A. New matrix formulas for Laguerre matrix polynomials. *J. Class. Anal.* **2013**, *3*, 59–67. [CrossRef]
4. Cortęs, J.C.; Jǫdar, L.; Company, R.; Villafuerte, L. Laguerre random polynomials: Definition, differential and statistical properties. *Util. Math.* **2015**, *98*, 283–293.

5.   Jódar, L.; Sastre, J. On the Laguerre matrix polynomials. *Util. Math.* **1998**, *53*, 37–48.

6.   Kargina, L.; Kurta, V. Modified Laguerre matrix polynomials. *Filomat* **2014**, *28*, 2069–2076. [CrossRef]

7.   Shehata, A. Some relations on Laguerre matrix polynomials. *Malays. J. Math. Sci.* **2015**, *9*, 443–462.

8.   Abdalla, M. Special matrix functions: Characteristics, achievements and future directions. *Linear Multilinear Algebra* **2018**. [CrossRef]

9.   Khan, S.; Hassan, N.A.M. 2-variables Laguerre matrix polynomials and Lie-algebraic techniques. *J. Phys. A* **2010**, *43*, 235204. [CrossRef]

10.   Yasmin, G.; Khan, S. Hermite matrix based polynomials of two variables and lie algebraic techniques. *Southeast Asian Bull. Math.* **2014**, *38*, 603–618.

11.   Aktaş, R.; Çekim, B.; Şahin, R. The matrix version for the multivariable Humbert polynomials. *Miskolc Math. Notes* **2012**, *13*, 197–208. [CrossRef]

12.   Bin-Saad, M.G. A note on two variable Laguerre matrix polynomials. *J. Associ. Arab Univ. Basic Appl. Sci.* **2017**, *24*, 271–276. [CrossRef]

13.   Bin-Saad, M.G.; Pathan, M.A. Pseudo laguerre matrix polynomials, operational identities and quasi-monomiality. *Adv. Linear Algebra Matrix Theory* **2018**, *8*, 87–95. [CrossRef]

14.   Çekim, B. *q*-Matrix polynomials in several variables. *Filomat* **2015**, *29*, 2059–2067. [CrossRef]

15.   Kahmmash, G.S. A study of a two variables Gegenbauer matrix polynomials and second order matrix partial differential equations. *Int. J. Math. Anal.* **2008**, *2*, 807–821.

16.   Khammash, G.S. On Hermite matrix polynomials of two variables. *J. Appl. Sci.* **2008**, *8*, 1221–1227.

17.   Singh, V.; Khan, M.; Khan, A. On a multivariable extension of Laguerre matrix ploynomials. *Electron. J. Math. Anal. Appl.* **2018**, *6*, 223–240.

18.   Abul-Dahab, M.A.; Abul-Ez, M.; Kishka, Z.; Constales, D. Reverse generalized Bessel matrix differential equation, polynomial solutions,and their properties. *Math. Methods Appl. Sci.* **2015**, *38*, 1005–1013. [CrossRef]

19.   Cortęs, J.C.; Jǫdar, L.; Sols, F.J.; Ku Carrillo, R. Infinite matrix products and the representation of the gamma matrix function. *Abstr. Appl. Anal.* **2015**, *2015*, 564287. [CrossRef]

20.   Jódar, L.; Cortés, J.C. On the hypergeometric matrix function. *J. Comput. Appl. Math.* **1998**, *99*, 205–217. [CrossRef]

21.   Abdalla, M.; Abd-Elmageed, H.; Abul-Ez, M.; Kishka, Z. Operational formulae of the multivariable hypergeometric matrix functions and related matrix polynomials. *Gen. Lett. Math.* **2017**, *3*, 81–90. [CrossRef]

22.   Khan, M.A.; Khan, A.H.; Ahmad, N. A note on Hermite polynomials of several variables. *Appl. Math. Comput.* **2012**, *218*, 6385–6390.

23.   Bakhet, A.; Jiao, Y.; He, F.L. On the Wright hypergeometric matrix functions and their fractional calculus. *Integral Transform. Spec. Funct.* **2019**, *30*, 138–156. [CrossRef]

24.   Defez, E.; Jódar, L. Some applications of the Hermite matrix polynomials series expansions. *J. Comput. Appl. Math.* **1998**, *99*, 105–117. [CrossRef]

25.   Abdalla, M.; Bakhet, A. Extended Gauss hypergeometric matrix functions. *Iran. J. Sci. Technol. Trans. Sci.* **2018**, *42*, 1465–1470. [CrossRef]

© 2019 by the authors. Licensee MDPI, Basel, Switzerland. This article is an open access article distributed under the terms and conditions of the Creative Commons Attribution (CC BY) license (http://creativecommons.org/licenses/by/4.0/).

Article

Some Symmetric Identities for Degenerate Carlitz-type (p,q)-Euler Numbers and Polynomials

Kyung-Won Hwang [1] and Cheon Seoung Ryoo [2],*

[1]   Department of Mathematics, Dong-A University, Busan 604-714, Korea; khwang@dau.ac.kr
[2]   Department of Mathematics, Hannam University, Daejeon 34430, Korea
*   Correspondence: ryoocs@hnu.kr

Received: 30 May 2019; Accepted: 20 June 2019; Published: 24 June 2019

Abstract: In this paper we define the degenerate Carlitz-type (p,q)-Euler polynomials by generalizing the degenerate Euler numbers and polynomials, degenerate Carlitz-type q-Euler numbers and polynomials. We also give some theorems and exact formulas, which have a connection to degenerate Carlitz-type (p,q)-Euler numbers and polynomials.

Keywords: degenerate Euler numbers and polynomials; degenerate q-Euler numbers and polynomials; degenerate Carlitz-type (p,q)-Euler numbers and polynomials

MSC: 11B68; 11S40; 11S80

---

1. Introduction

Many researchers have studied about the degenerate Bernoulli numbers and polynomials, degenerate Euler numbers and polynomials, degenerate Genocchi numbers and polynomials, degenerate tangent numbers and polynomials (see [1–7]). Recently, some generalizations of the Bernoulli numbers and polynomials, Euler numbers and polynomials, Genocchi numbers and polynomials, tangent numbers and polynomials are provided (see [6,8–13]). In this paper we define the degenerate Carlitz-type (p,q)-Euler polynomials and numbers and study some theories of the degenerate Carlitz-type (p,q)-Euler numbers and polynomials.

Throughout this paper, we use the notations below: $\mathbb{N}$ denotes the set of natural numbers, $\mathbb{Z}$ denotes the set of integers, $\mathbb{Z}_+ = \mathbb{N} \cup \{0\}$ denotes the set of nonnegative integers. We remind that the classical degenerate Euler numbers $\mathcal{E}_n(\lambda)$ and Euler polynomials $\mathcal{E}_n(x,\lambda)$, which are defined by generating functions like (1), and (2) (see [1,2])

$$\frac{2}{(1+\lambda t)^{\frac{1}{\lambda}}+1} - \sum_{n=0}^{\infty} \mathcal{E}_n(\lambda)\frac{t^n}{n!}, \tag{1}$$

and

$$\frac{2}{(1+\lambda t)^{\frac{1}{\lambda}}+1}(1+\lambda t)^{\frac{x}{\lambda}} = \sum_{n=0}^{\infty} \mathcal{E}_n(x,\lambda)\frac{t^n}{n!}, \tag{2}$$

respectively.

Carlitz [1] introduced some theories of the degenerate Euler numbers and polynomials. We recall that well-known Stirling numbers of the first kind $S_1(n,k)$ and the second kind $S_2(n,k)$ are defined by this (see [2,7,14])

$$(x)_n = \sum_{k=0}^{n} S_1(n,k)x^k \text{ and } x^n = \sum_{k=0}^{n} S_2(n,k)(x)_k,$$

respectively. Here $(x)_n = x(x-1)\cdots(x-n+1)$. The numbers $S_2(n,m)$ is like this

$$\sum_{n=m}^{\infty} S_2(n,m)\frac{t^n}{n!} = \frac{(e^t-1)^m}{m!}.$$

We also have

$$\sum_{n=m}^{\infty} S_1(n,m)\frac{t^n}{n!} = \frac{(\log(1+t))^m}{m!}.$$

The generalized falling factorial $(x|\lambda)_n$ with increment λ is defined by

$$(x|\lambda)_n = \prod_{k=0}^{n-1}(x-\lambda k)$$

for positive integer n, with $(x|\lambda)_0 = 1$; as we know,

$$(x|\lambda)_n = \sum_{k=0}^{n} S_1(n,k)\lambda^{n-k}x^k.$$

$(x|\lambda)_n = \lambda^n(\lambda^{-1}x|1)_n$ for $\lambda \neq 0$. Clearly $(x|0)_n = x^n$. The binomial theorem for a variable x is

$$(1+\lambda t)^{x/\lambda} = \sum_{n=0}^{\infty}(x|\lambda)_n\frac{t^n}{n!}.$$

The (p,q)-number is defined as

$$[n]_{p,q} = \frac{p^n - q^n}{p - q} = p^{n-1} + p^{n-2}q + p^{n-3}q^2 + \cdots + p^2q^{n-3} + pq^{n-2} + q^{n-1}.$$

We begin by reminding the Carlitz-type (p,q)-Euler numbers and polynomials (see [9–11]).

Definition 1. *For $0 < q < p \leq 1$ and $h \in \mathbb{Z}$, the Carlitz-type (p,q)-Euler polynomials $E_{n,p,q}(x)$ and (h,p,q)-Euler polynomials $E_{n,p,q}^{(h)}(x)$ are defined like this*

$$\sum_{n=0}^{\infty} E_{n,p,q}(x)\frac{t^n}{n!} = [2]_q \sum_{m=0}^{\infty}(-1)^m q^m e^{[m+x]_{p,q}t},$$

$$\sum_{n=0}^{\infty} E_{n,p,q}^{(h)}(x)\frac{t^n}{n!} = [2]_q \sum_{m=0}^{\infty}(-1)^m q^m p^{hm} e^{[m+x]_{p,q}t},$$

(3)

respectively (see [9–11]).

Now we make the degenerate Carlitz-type (p,q)-Euler number $\mathcal{E}_{n,p,q}(\lambda)$ and (p,q)-Euler polynomials $\mathcal{E}_{n,p,q}(x,\lambda)$. In the next section, we introduce the degenerate Carlitz-type (p,q)-Euler numbers and polynomials. We will study some their properties after introduction.

2. Degenerate Carlitz-Type (p,q)-Euler Polynomials

In this section, we define the degenerate Carlitz-type (p,q)-Euler numbers and polynomials and make some of their properties.

Definition 2. *For $0 < q < p \le 1$, the degenerate Carlitz-type (p,q)-Euler numbers $\mathcal{E}_{n,p,q}(\lambda)$ and polynomials $\mathcal{E}_{n,p,q}(x,\lambda)$ are related to the generating functions*

$$F_{p,q}(t,\lambda) = \sum_{n=0}^{\infty} \mathcal{E}_{n,p,q}(\lambda) \frac{t^n}{n!} = [2]_q \sum_{m=0}^{\infty} (-1)^m q^m (1+\lambda t)^{\frac{[m]_{p,q}}{\lambda}}, \tag{4}$$

and

$$F_{p,q}(t,x,\lambda) = \sum_{n=0}^{\infty} \mathcal{E}_{n,p,q}(x,\lambda) \frac{t^n}{n!} = [2]_q \sum_{m=0}^{\infty} (-1)^m q^m (1+\lambda t)^{\frac{[m+x]_{p,q}}{\lambda}}, \tag{5}$$

respectively.

Let $p = 1$ in (4) and (5), we can get the degenerate Carlitz-type q-Euler number $\mathcal{E}_{n,q}(x,\lambda)$ and q-Euler polynomials $\mathcal{E}_{n,q}(x,\lambda)$ respectively. Obviously, if $p = 1$, then we have

$$\mathcal{E}_{n,p,q}(x,\lambda) = \mathcal{E}_{n,q}(x,\lambda), \quad \mathcal{E}_{n,p,q}(\lambda) = \mathcal{E}_{n,q}(\lambda).$$

When $p = 1$, we have

$$\lim_{q \to 1} \mathcal{E}_{n,p,q}(x,\lambda) = \mathcal{E}_n(x,\lambda), \quad \lim_{q \to 1} \mathcal{E}_{n,p,q}(\lambda) = \mathcal{E}_n(\lambda).$$

We see that

$$\begin{aligned}
(1+\lambda t)^{\frac{[x+y]_{p,q}}{\lambda}} &= e^{\frac{[x+y]_{p,q}}{\lambda} \log(1+\lambda t)} \\
&= \sum_{n=0}^{\infty} \left(\frac{[x+y]_{p,q}}{\lambda} \right)^n \frac{(\log(1+\lambda t))^n}{n!} \\
&= \sum_{n=0}^{\infty} \left(\sum_{m=0}^{n} S_1(n,m) \lambda^{n-m} [x+y]_{p,q}^m \right) \frac{t^n}{n!}.
\end{aligned} \tag{6}$$

By (5), it follows that

$$\begin{aligned}
&\sum_{n=0}^{\infty} \mathcal{E}_{n,p,q}(x,\lambda) \frac{t^n}{n!} \\
&= [2]_q \sum_{m=0}^{\infty} (-1)^m q^m (1+\lambda t)^{\frac{[m+x]_{p,q}}{\lambda}} \\
&= [2]_q \sum_{m=0}^{\infty} (-1)^m q^m \\
&\quad \times \sum_{n=0}^{\infty} \sum_{l=0}^{n} S_1(n,l) \lambda^{n-l} \frac{\sum_{j=0}^{l} \binom{l}{j} (-1)^j p^{(x+m)(l-j)} q^{(x+m)j}}{(p-q)^l} \frac{t^n}{n!} \\
&= \sum_{n=0}^{\infty} \left([2]_q \sum_{l=0}^{n} \sum_{j=0}^{l} \frac{S_1(n,l) \lambda^{n-l} \binom{l}{j} (-1)^j q^{xj} p^{x(l-j)}}{(p-q)^l} \frac{1}{1+q^{j+1} p^{l-j}} \right) \frac{t^n}{n!}.
\end{aligned} \tag{7}$$

By comparing the coefficients $\frac{t^n}{n!}$ in the above equation, we have the following theorem.

Theorem 1. *For $0 < q < p \leq 1$ and $n \in \mathbb{Z}_+$, we have*

$$
\begin{aligned}
\mathcal{E}_{n,p,q}(x,\lambda) &= [2]_q \sum_{l=0}^{n} \sum_{j=0}^{l} \frac{S_1(n,l)\lambda^{n-l}\binom{l}{j}(-1)^j q^{xj} p^{x(l-j)}}{(p-q)^l} \frac{1}{1+q^{j+1}p^{l-j}} \\
&= [2]_q \sum_{m=0}^{\infty} \sum_{l=0}^{n} S_1(n,l)\lambda^{n-l}(-1)^m q^m [x+m]_{p,q}^l, \\
\mathcal{E}_{n,p,q}(\lambda) &= [2]_q \sum_{l=0}^{n} \sum_{j=0}^{l} \frac{S_1(n,l)\lambda^{n-l}\binom{l}{j}(-1)^j}{(p-q)^l} \frac{1}{1+q^{j+1}p^{l-j}} \\
&= [2]_q \sum_{m=0}^{\infty} \sum_{l=0}^{n} S_1(n,l)\lambda^{n-l}(-1)^m q^m [m]_{p,q}^l.
\end{aligned}
$$

We make the degenerate Carlitz-type (p,q)-Euler number $\mathcal{E}_{n,p,q}(\lambda)$. Some cases are

$$
\begin{aligned}
\mathcal{E}_{0,p,q}(\lambda) &= 1, \\
\mathcal{E}_{1,p,q}(\lambda) &= \frac{[2]_q}{(p-q)(1+pq)} - \frac{[2]_q}{(p-q)(1+q^2)}, \\
\mathcal{E}_{2,p,q}(\lambda) &= -\frac{[2]_q \lambda}{(p-q)(1+pq)} + \frac{[2]_q}{(p-q)^2(1+p^2q)} + \frac{[2]_q \lambda}{(p-q)(1+q^2)} \\
&\quad - \frac{2[2]_q}{(p-q)^2(1+pq^2)} + \frac{[2]_q}{(p-q)^2(1+q^3)}, \\
\mathcal{E}_{3,p,q}(\lambda) &= \frac{2[2]_q \lambda^2}{(p-q)(1+pq)} - \frac{3[2]_q \lambda}{(p-q)^2(1+p^2q)} + \frac{[2]_q}{(p-q)^3(1+p^3q)} \\
&\quad - \frac{2[2]_q \lambda^2}{(p-q)(1+q^2)} + \frac{6[2]_q \lambda}{(p-q)^2(1+pq^2)} - \frac{3[2]_q}{(p-q)^3(1+p^2q^2)} \\
&\quad - \frac{3[2]_q \lambda}{(p-q)^2(1+q^3)} + \frac{3[2]_q}{(p-q)^3(1+pq^3)} - \frac{[2]_q}{(p-q)^3(1+q^4)}.
\end{aligned}
$$

We use t instead of $\dfrac{e^{\lambda t}-1}{\lambda}$ in (5), we have

$$
\begin{aligned}
\sum_{m=0}^{\infty} E_{m,p,q}(x)\frac{t^m}{m!} &= \sum_{n=0}^{\infty} \mathcal{E}_{n,p,q}(x,\lambda)\left(\frac{e^{\lambda t}-1}{\lambda}\right)^n \frac{1}{n!} \\
&= \sum_{n=0}^{\infty} \mathcal{E}_{n,p,q}(x,\lambda)\lambda^{-n} \sum_{m=n}^{\infty} S_2(m,n)\lambda^m \frac{t^m}{m!} \\
&= \sum_{m=0}^{\infty} \left(\sum_{n=0}^{m} \mathcal{E}_{n,p,q}(x,\lambda)\lambda^{m-n} S_2(m,n)\right) \frac{t^m}{m!}.
\end{aligned}
\tag{8}
$$

Thus we have the following theorem.

Theorem 2. *For $m \in \mathbb{Z}_+$, we have*

$$
E_{m,p,q}(x) = \sum_{n=0}^{m} \mathcal{E}_{n,p,q}(x,\lambda)\lambda^{m-n} S_2(m,n).
$$

Use t instead of $\log(1 + \lambda t)^{1/\lambda}$ in (3), we have

$$\sum_{n=0}^{\infty} E_{n,p,q}(x) \left(\log(1 + \lambda t)^{1/\lambda} \right)^n \frac{1}{n!}$$

$$= [2]_q \sum_{m=0}^{\infty} (-1)^m q^m (1 + \lambda t)^{\frac{[m+x]_{p,q}}{\lambda}} \tag{9}$$

$$= \sum_{m=0}^{\infty} \mathcal{E}_{m,p,q}(x, \lambda) \frac{t^m}{m!},$$

and

$$\sum_{n=0}^{\infty} E_{n,p,q}(x) \left(\log(1 + \lambda t)^{1/\lambda} \right)^n \frac{1}{n!}$$

$$= \sum_{m=0}^{\infty} \left(\sum_{n=0}^{m} E_{n,p,q}(x) \lambda^{m-n} S_1(m, n) \right) \frac{t^m}{m!}. \tag{10}$$

Thus we have the below theorem from (9) and (10).

Theorem 3. *For* $m \in \mathbb{Z}_+$, *we have*

$$\mathcal{E}_{m,p,q}(x, \lambda) = \sum_{n=0}^{m} E_{n,p,q}(x) \lambda^{m-n} S_1(m, n).$$

We have the degenerate Carlitz-type (p, q)-Euler polynomials $\mathcal{E}_{n,p,q}(x, \lambda)$. some cases are

$$\mathcal{E}_{0,p,q}(x, \lambda) = 1,$$

$$\mathcal{E}_{1,p,q}(x, \lambda) = \frac{[2]_q p^x}{(p-q)(1+pq)} - \frac{[2]_q q^x}{(p-q)(1+q^2)},$$

$$\mathcal{E}_{2,p,q}(x, \lambda) = -\frac{[2]_q \lambda p^x}{(p-q)(1+pq)} + \frac{[2]_q p^{2x}}{(p-q)^2(1+p^2q)} + \frac{[2]_q \lambda q^x}{(p-q)(1+q^2)}$$

$$- \frac{2[2]_q p^x q^x}{(p-q)^2(1+pq^2)} + \frac{[2]_q q^{2x}}{(p-q)^2(1+q^3)},$$

$$\mathcal{E}_{3,p,q}(x, \lambda) = \frac{2[2]_q \lambda^2 p^x}{(p-q)(1+pq)} - \frac{3[2]_q \lambda p^{2x}}{(p-q)^2(1+p^2q)} + \frac{[2]_q p^{3x}}{(p-q)^3(1+p^3q)}$$

$$- \frac{2[2]_q \lambda^2 q^x}{(p-q)(1+q^2)} + \frac{6[2]_q \lambda p^x q^x}{(p-q)^2(1+pq^2)} - \frac{3[2]_q p^{2x} q^x}{(p-q)^3(1+p^2q^2)}$$

$$- \frac{3[2]_q \lambda q^{2x}}{(p-q)^2(1+q^3)} + \frac{3[2]_q p^x q^{2x}}{(p-q)^3(1+pq^3)} - \frac{[2]_q q^{3x}}{(p-q)^3(1+q^4)}.$$

We introduce a (p, q)-analogue of the generalized falling factorial $(x|\lambda)_n$ with increment λ. The generalized (p, q)-falling factorial $([x]_{p,q}|\lambda)_n$ with increment λ is defined by

$$([x]_{p,q}|\lambda)_n = \prod_{k=0}^{n-1} ([x]_{p,q} - \lambda k)$$

for positive integer n, where $([x]_{p,q}|\lambda)_0 = 1$.

By (4) and (5), we get

$$- [2]_q (-1)^n q^n \sum_{l=0}^{\infty} (-1)^l q^l (1 + \lambda t)^{\frac{[l+n]_{p,q}}{\lambda}}$$

$$+ [2]_q \sum_{l=0}^{\infty} (-1)^l q^l (1 + \lambda t)^{\frac{[l+n]_{p,q}}{\lambda}}$$

$$= [2]_q \sum_{l=0}^{n-1} (-1)^l q^l (1 + \lambda t)^{\frac{[l]_{p,q}}{\lambda}} .$$

Hence we have

$$(-1)^{n+1} q^n \sum_{m=0}^{\infty} \mathcal{E}_{m,p,q}(n, \lambda) \frac{t^m}{m!} + \sum_{m=0}^{\infty} \mathcal{E}_{m,p,q}(\lambda) \frac{t^m}{m!}$$

$$= \sum_{m=0}^{\infty} \left([2]_q \sum_{l=0}^{n-1} (-1)^l q^l ([l]_{p,q}|\lambda)_m \right) \frac{t^m}{m!}. \tag{11}$$

By comparing the coefficients of $\frac{t^m}{m!}$ on both sides of (11), we have the following theorem.

Theorem 4. *For $n \in \mathbb{Z}_+$, we have*

$$\sum_{l=0}^{n-1} (-1)^l q^l ([l]_{p,q}|\lambda)_m = \frac{(-1)^{n+1} q^n \mathcal{E}_{m,p,q}(n, \lambda) + \mathcal{E}_{m,p,q}(\lambda)}{[2]_q}.$$

We get that

$$(1 + \lambda t)^{\frac{[x+y]_{p,q}}{\lambda}}$$

$$= (1 + \lambda t)^{\frac{p^y [x]_{p,q}}{\lambda}} (1 + \lambda t)^{\frac{q^x [y]_{p,q}}{\lambda}}$$

$$= \sum_{m=0}^{\infty} (p^y [x]_{p,q}|\lambda)_m \frac{t^m}{m!} e^{\log(1+\lambda t) \frac{q^x [y]_{p,q}}{\lambda}}$$

$$= \sum_{m=0}^{\infty} (p^y [x]_{p,q}|\lambda)_m \frac{t^m}{m!} \sum_{l=0}^{\infty} \left(\frac{q^x [y]_{p,q}}{\lambda} \right)^l \frac{\log(1 + \lambda t)^l}{l!} \tag{12}$$

$$= \sum_{m=0}^{\infty} (p^y [x]_{p,q}|\lambda)_m \frac{t^m}{m!} \sum_{l=0}^{\infty} \left(\frac{q^x [y]_{p,q}}{\lambda} \right)^l \sum_{k=l}^{\infty} S_1(k, l) \lambda^k \frac{t^k}{k!}$$

$$= \sum_{n=0}^{\infty} \left(\sum_{k=0}^{n} \sum_{l=0}^{k} \binom{n}{k} (p^y [x]_{p,q}|\lambda)_{n-k} \lambda^{k-l} q^{xl} [y]_{p,q}^l S_1(k, l) \right) \frac{t^n}{n!}.$$

By (5) and (12), we get

$$\sum_{n=0}^{\infty} \mathcal{E}_{n,p,q\zeta}(x,\lambda)\frac{t^n}{n!}$$

$$= [2]_q \sum_{m=0}^{\infty} (-1)^m q^m (1+\lambda t)^{\frac{[m+x]_{p,q}}{\lambda}}$$

$$= [2]_q \sum_{m=0}^{\infty} (-1)^m q^m \sum_{n=0}^{\infty} \left(\sum_{k=0}^{n} \sum_{l=0}^{k} \binom{n}{k} (p^m [x]_{p,q}|\lambda)_{n-k} \lambda^{k-l} q^{xl} [m]_{p,q}^l S_1(k,l) \right) \frac{t^n}{n!}$$

$$= \sum_{n=0}^{\infty} \left([2]_q \sum_{m=0}^{\infty} \sum_{k=0}^{n} \sum_{l=0}^{k} \binom{n}{k} (-1)^m q^m (p^m [x]_{p,q}|\lambda)_{n-k} \lambda^{k-l} q^{xl} S_1(k,l) \right) \frac{t^n}{n!}.$$

By comparing the coefficients of $\frac{t^n}{n!}$ in the above equation, we have the theorem below.

Theorem 5. *For $0 < q < p \le 1$ and $n \in \mathbb{Z}_+$, we have*

$$\mathcal{E}_{n,p,q}(x,\lambda) = [2]_q \sum_{m=0}^{\infty} \sum_{k=0}^{n} \sum_{l=0}^{k} \binom{n}{k} (-1)^m q^m (p^m [x]_{p,q}|\lambda)_{n-k} \lambda^{k-l} q^{xl} S_1(k,l).$$

3. Symmetric Properties about Degenerate Carlitz-Type (p,q)-Euler Numbers and Polynomials

In this section, we are going to get the main results of degenerate Carlitz-type (p,q)-Euler numbers and polynomials. We also make some symmetric identities for degenerate Carlitz-type (p,q)-Euler numbers and polynomials. Let w_1 and w_2 be odd positive integers. Remind that $[xy]_{p,q} = [x]_{p^y,q^y}[y]_{p,q}$ for any $x, y \in \mathbb{C}$.

By using $w_1 x + \frac{w_1 i}{w_2}$ instead of x in Definition 2, use p by p^{w_2}, use q by q^{w_2} and use λ by $\frac{\lambda}{[w_2]_{p,q}}$, respectively, we can get

$$\sum_{n=0}^{\infty} \left([2]_{q^{w_1}} [w_2]_{p,q}^n \sum_{i=0}^{w_2-1} (-1)^i q^{w_1 i} \mathcal{E}_{n,p^{w_2},q^{w_2}} \left(w_1 x + \frac{w_1 i}{w_2}, \frac{\lambda}{[w_2]_{p,q}} \right) \right) \frac{t^n}{n!}$$

$$= [2]_{q^{w_1}} \sum_{i=0}^{w_2-1} (-1)^i q^{w_1 i} \sum_{n=0}^{\infty} \mathcal{E}_{n,p^{w_2},q^{w_2}} \left(w_1 x + \frac{w_1 i}{w_2}, \frac{\lambda}{[w_2]_{p,q}} \right) \frac{([w_2]_{p,q}t)^n}{n!}$$

$$= [2]_{q^{w_1}} \sum_{i=0}^{w_2-1} (-1)^i q^{w_1 i} [2]_{q^{w_2}} \sum_{n=0}^{\infty} (-1)^n q^{w_2 n}$$

$$\times \left(1 + \frac{\lambda}{[w_2]_{p,q}} [w_2]_{p,q} t \right)^{\frac{[w_1 x + \frac{w_1 i}{w_2} + n]_{p^{w_2},q^{w_2}}}{\frac{\lambda}{[w_2]_{p,q}}}}$$

$$= [2]_{q^{w_1}} \sum_{i=0}^{w_2-1} (-1)^i q^{w_1 i} [2]_{q^{w_2}} \sum_{n=0}^{\infty} (-1)^n q^{w_2 n}$$

$$\times (1 + \lambda t)^{\frac{[w_1 w_2 x + w_1 i + n w_2]_{p,q}}{\lambda}}.$$

Since for any non-negative integer n and odd positive integer w_1, there is the unique non-negative integer r such that $n = w_1 r + j$ with $0 \leq j \leq w_1 - 1$. So this can be written as

$$[2]_{q^{w_1}} [2]_{q^{w_2}} \sum_{i=0}^{w_2-1} (-1)^i q^{w_1 i} \sum_{n=0}^{\infty} (-1)^n q^{w_2 n}$$
$$\times (1 + \lambda t)^{\frac{[w_1 w_2 x + w_1 i + n w_2]_{p,q}}{\lambda}}.$$

$$= [2]_{q^{w_1}} [2]_{q^{w_2}} \sum_{i=0}^{w_2-1} (-1)^i q^{w_1 i} \sum_{\substack{w_1 r + j = 0 \\ 0 \leq j \leq w_1 - 1}}^{\infty} (-1)^{w_1 r + j} q^{w_2(w_1 r + j)}$$
$$\times (1 + \lambda t)^{\frac{[w_1 w_2 x + w_1 i + (w_1 r + j) w_2]_{p,q}}{\lambda}}.$$

$$= [2]_{q^{w_1}} [2]_{q^{w_2}} \sum_{i=0}^{w_2-1} (-1)^i q^{w_1 i} \sum_{j=0}^{w_1-1} \sum_{r=0}^{\infty} (-1)^{w_1 r} (-1)^j q^{w_2 w_1 r} q^{w_2 j}$$
$$\times (1 + \lambda t)^{\frac{[w_1 w_2 x + w_1 i + w_1 w_2 r + w_2 j]_{p,q}}{\lambda}}.$$

$$= [2]_{q^{w_1}} [2]_{q^{w_2}} \sum_{i=0}^{w_2-1} \sum_{j=0}^{w_1-1} \sum_{r=0}^{\infty} (-1)^i (-1)^r (-1)^j q^{w_1 i} q^{w_2 w_1 r} q^{w_2 j}$$
$$\times (1 + \lambda t)^{\frac{[w_1 w_2 x + w_1 i + w_1 w_2 r + w_2 j]_{p,q}}{\lambda}}.$$

We have the below formula using the above formula

$$\sum_{n=0}^{\infty} \left([2]_{q^{w_2}} [w_2]_{p,q}^n \sum_{i=0}^{w_2-1} (-1)^i q^{w_1 i} \mathcal{E}_{n, p^{w_2}, q^{w_2}} \left(w_1 x + \frac{w_1 i}{w_2}, \frac{\lambda}{[w_2]_{p,q}} \right) \right) \frac{t^n}{n!}$$
$$= [2]_{q^{w_1}} [2]_{q^{w_2}} \sum_{i=0}^{w_2-1} \sum_{j=0}^{w_1-1} \sum_{r=0}^{\infty} (-1)^i (-1)^r (-1)^j q^{w_1 i} q^{w_2 w_1 r} q^{w_2 j} \tag{13}$$
$$\times (1 + \lambda t)^{\frac{[w_1 w_2 x + w_1 i + w_1 w_2 r + w_2 j]_{p,q}}{\lambda}}.$$

From a similar approach, we can have that

$$\sum_{n=0}^{\infty} \left([2]_{q^{w_2}} [w_1]_{p,q}^n \sum_{i=0}^{w_1-1} (-1)^i q^{w_2 i} \mathcal{E}_{n, p^{w_1}, q^{w_1}} \left(w_2 x + \frac{w_2 i}{w_1}, \frac{\lambda}{[w_1]_{p,q}} \right) \right) \frac{t^n}{n!}$$
$$= [2]_{q^{w_1}} [2]_{q^{w_2}} \sum_{i=0}^{w_1-1} \sum_{j=0}^{w_2-1} \sum_{r=0}^{\infty} (-1)^i (-1)^r (-1)^j q^{w_2 i} q^{w_1 w_1 r} q^{w_1 j} \tag{14}$$
$$\times (1 + \lambda t)^{\frac{[w_1 w_2 x + w_2 i + w_1 w_2 r + w_1 j]_{p,q}}{\lambda}}.$$

Thus, we have the following theorem from (13) and (14).

Theorem 6. *Let w_1 and w_2 be odd positive integers. Then one has*

$$[2]_{q^{w_1}}[w_2]_{p,q}^n \sum_{i=0}^{w_2-1} (-1)^i q^{w_1 i} \mathcal{E}_{n,p^{w_2},q^{w_2}}\left(w_1 x + \frac{w_1 i}{w_2}, \frac{\lambda}{[w_2]_{p,q}}\right)$$

$$= [2]_{q^{w_2}}[w_1]_{p,q}^n \sum_{j=0}^{w_1-1} (-1)^j q^{w_2 j} \mathcal{E}_{n,p^{w_1},q^{w_1}}\left(w_2 x + \frac{w_2 j}{w_1}, \frac{\lambda}{[w_1]_{p,q}}\right).$$

Letting $\lambda \to 0$ in Theorem 6, we can immediately obtain the symmetric identities for Carlitz-type (p,q)-Euler polynomials (see [10])

$$[2]_{q^{w_1}}[w_2]_{p,q}^n \sum_{i=0}^{w_2-1} (-1)^i q^{w_1 i} E_{n,p^{w_2},q^{w_2}}\left(w_1 x + \frac{w_1 i}{w_2}\right)$$

$$= [2]_{q^{w_2}}[w_1]_{p,q}^n \sum_{j=0}^{w_1-1} (-1)^j q^{w_2 j} E_{n,p^{w_1},q^{w_1}}\left(w_2 x + \frac{w_2 j}{w_1}\right).$$

It follows that we show some special cases of Theorem 6. Let $w_2 = 1$ in Theorem 6, we have the multiplication theorem for the degenerate Carlitz-type (p,q)-Euler polynomials.

Corollary 1. *Let w_1 be odd positive integer. Then*

$$\mathcal{E}_{n,p,q}(x,\lambda) = \frac{[2]_q[w_1]_{p,q}^n}{[2]_{q^{w_1}}} \sum_{j=0}^{w_1-1} (-1)^j q^j \mathcal{E}_{n,p^{w_1},q^{w_1}}\left(\frac{x+j}{w_1}, \frac{\lambda}{[w_1]_{p,q}}\right). \tag{15}$$

Let $p = 1$ in (15). This leads to the multiplication theorem about the degenerate Carlitz-type q-Euler polynomials

$$\mathcal{E}_{n,q}(x,\lambda) = \frac{[2]_q[w_1]_q^n}{[2]_{q^{w_1}}} \sum_{j=0}^{w_1-1} (-1)^j q^j \mathcal{E}_{n,q^{w_1}}\left(\frac{x+j}{w_1}, \frac{\lambda}{[w_1]_q}\right). \tag{16}$$

Giving $q \to 1$ in (16) induce to the multiplication theorem about the degenerate Euler polynomials

$$\mathcal{E}_n(x,\lambda) = w_1^n \sum_{j=0}^{w_1-1} (-1)^j \mathcal{E}_n\left(\frac{x+i}{w_1}, \frac{\lambda}{w_1}\right). \tag{17}$$

If λ approaches to 0 in (17), this leads to the multiplication theorem about the Euler polynomials(see [15])

$$E_n(x) = w_1^n \sum_{j=0}^{w_1-1} (-1)^j E_n\left(\frac{x+i}{w_1}\right).$$

Let $x = 0$ in Theorem 6, then we have the following corollary.

Corollary 2. *Let w_1 and w_2 be odd positive integers. Then it has*

$$[2]_{q^{w_1}}[w_2]_{p,q}^n \sum_{i=0}^{w_2-1} (-1)^i q^{w_1 i} \mathcal{E}_{n,p^{w_2},q^{w_2}}\left(\frac{w_1 i}{w_2}, \frac{\lambda}{[w_2]_{p,q}}\right)$$

$$= [2]_{q^{w_2}}[w_1]_{p,q}^n \sum_{j=0}^{w_1-1} (-1)^j q^{w_2 j} \mathcal{E}_{n,p^{w_1},q^{w_1}}\left(\frac{w_2 j}{w_1}, \frac{\lambda}{[w_1]_{p,q}}\right).$$

By Theorem 3 and Corollary 2, we have the below theorem.

Theorem 7. *Let w_1 and w_2 be odd positive integers. Then*

$$\sum_{l=0}^{n} S_1(n,l)\lambda^{n-l}[w_2]_{p,q}^{l}[2]_{q^{w_1}} \sum_{i=0}^{w_2-1}(-1)^i q^{w_1 i} E_{l,p^{w_2},q^{w_2}}\left(\frac{w_1}{w_2}i\right)$$

$$=\sum_{l=0}^{n} S_1(n,l)\lambda^{n-l}[w_1]_{p,q}^{l}[2]_{q^{w_2}} \sum_{j=0}^{w_1-1}(-1)^j q^{w_2 j} E_{l,p^{w_1},q^{w_1}}\left(\frac{w_2}{w_1}j\right).$$

We get another result by applying the addition theorem about the Carlitz-type (p,q)-Euler polynomials $E_{n,p,q}(x)$.

Theorem 8. *Let w_1 and w_2 be odd positive integers. Then we have*

$$\sum_{l=0}^{n}\sum_{k=0}^{l}\binom{l}{k}S_1(n,l)\lambda^{n-l}p^{w_1 w_2 xk}[2]_{q^{w_1}}[w_1]_{p,q}^{k}[w_2]_{p,q}^{l-k}E_{l-k,p^{w_2},q^{w_2}}^{(k)}(w_1 x)S_{l,k,p^{w_1},q^{w_1}}(w_2)$$

$$=\sum_{l=0}^{n}\sum_{k=0}^{l}\binom{l}{k}S_1(n,l)\lambda^{n-l}p^{w_1 w_2 xk}[2]_{q^{w_2}}[w_2]_{p,q}^{k}[w_1]_{p,q}^{l-k}E_{l-k,p^{w_1},q^{w_1}}^{(k)}(w_2 x)S_{l,k,p^{w_2},q^{w_2}}(w_1),$$

where $S_{l,k,p,q}(w_1) = \sum_{i=0}^{w_1-1}(-1)^i q^{(l-k+1)i}[i]_{p,q}^{k}$ is called as the (p,q)-sums of powers.

Proof. From (3), Theorems 3 and 6, we have

$$[2]_{q^{w_1}}[w_2]_{p,q}^{n}\sum_{i=0}^{w_2-1}(-1)^i q^{w_1 i}\mathcal{E}_{n,p^{w_2},q^{w_2}}\left(w_1 x+\frac{w_1 i}{w_2},\frac{\lambda}{[w_2]_{p,q}}\right)$$

$$=[2]_{q^{w_1}}[w_2]_{p,q}^{n}\sum_{i=0}^{w_2-1}(-1)^i q^{w_1 i}\sum_{l=0}^{n}E_{l,p^{w_2},q^{w_2}}\left(w_1 x+\frac{w_1 i}{w_2}\right)\left(\frac{\lambda}{[w_2]_{p,q}}\right)^{n-l}S_1(n,l)$$

$$=[2]_{q^{w_1}}\sum_{l=0}^{n}S_1(n,l)\lambda^{n-l}[w_2]_{p,q}^{l}\sum_{i=0}^{w_2-1}(-1)^i q^{w_1 i}\sum_{k=0}^{l}q^{w_1(l-k)i}p^{w_1 w_2 xk}$$

$$\times E_{l-k,p^{w_2},q^{w_2}}^{(k)}(w_1 x)\left(\frac{[w_1]_{p,q}}{[w_2]_{p,q}}\right)^{k}[i]_{p^{w_1},q^{w_1}}^{k}$$

$$=[2]_{q^{w_1}}\sum_{l=0}^{n}S_1(n,l)\lambda^{n-l}\sum_{k=0}^{l}\binom{l}{k}p^{w_1 w_2 xk}[w_1]_{p,q}^{k}[w_2]_{p,q}^{l-k}p^{w_1 w_2 xl}E_{l-k,p^{w_2},q^{w_2}}^{(k)}(w_1 x)$$

$$\times\sum_{i=0}^{w_2-1}(-1)^i q^{w_1 i}q^{(l-k)w_1 i}[i]_{p^{w_1},q^{w_1}}^{k}.$$

Therefore, we induce that

$$[2]_{q^{w_1}}[w_2]_{p,q}^{n}\sum_{i=0}^{w_2-1}(-1)^i q^{w_1 i}\mathcal{E}_{n,p^{w_2},q^{w_2}}\left(w_1 x+\frac{w_1 i}{w_2},\frac{\lambda}{[w_2]_{p,q}}\right)$$

$$=\sum_{l=0}^{n}\sum_{k=0}^{l}\binom{l}{k}S_1(n,l)\lambda^{n-l}p^{w_1 w_2 xk}[2]_{q^{w_1}}[w_1]_{p,q}^{k}[w_2]_{p,q}^{l-k}p^{w_1 w_2 xl} \tag{18}$$

$$\times E_{l-k,p^{w_2},q^{w_2}}^{(k)}(w_1 x)S_{l,k,p^{w_1},q^{w_1}}(w_2),$$

and

$$[2]_{q^{w_2}} [w_1]_{p,q}^n \sum_{j=0}^{w_1-1} (-1)^j q^{w_2 j} \mathcal{E}_{n,p^{w_1},q^{w_1}} \left(w_2 x + \frac{w_2 j}{w_1}, \frac{\lambda}{[w_1]_{p,q}} \right)$$

$$= \sum_{l=0}^{n} \sum_{k=0}^{l} \binom{l}{k} S_1(n,l) \lambda^{n-l} p^{w_1 w_2 xk} [2]_{q^{w_2}} [w_2]_{p,q}^k [w_1]_{p,q}^{l-k} \tag{19}$$

$$\times E_{l-k,p^{w_1},q^{w_1}}^{(k)} (w_2 x) S_{l,k,p^{w_2},q^{w_2}} (w_1).$$

By (18) and (19), we make the desired symmetric identity. $\square$

Author Contributions: All authors contributed equally in writing this article. All authors read and approved the final manuscript.

Funding: This work was supported by the Dong-A university research fund.

Acknowledgments: The authors would like to thank the referees for their valuable comments, which improved the original manuscript in its present form.

Conflicts of Interest: The authors declare no conflict of interest.

References

1. Carlitz, L. Degenerate Stiling, Bernoulli and Eulerian numbers. *Utilitas Math.* **1979**, *15*, 51–88.
2. Cenkci, M.; Howard, F.T. Notes on degenerate numbers. *Discrete Math.* **2007**, *307*, 2359–2375. [CrossRef]
3. Howard, F.T. Degenerate weighted Stirling numbers. *Discrete Math.* **1985**, *57*, 45–58. [CrossRef]
4. Howard, F.T. Explicit formulas for degenerate Bernoulli numbers. *Discrete Math.* **1996**, *162*, 175–185. [CrossRef]
5. Qi, F.; Dolgy, D.V.; Kim, T.; Ryoo, C.S. On the partially degenerate Bernoulli polynomials of the first kind. *Glob. J. Pure Appl. Math.* **2015**, *11*, 2407–2412.
6. Ryoo, C.S. On degenerate Carlitz-type (h,q)-tangent numbers and polynomials. *J. Algebra Appl. Math.* **2018**, *16*, 119–130.
7. Ryoo, C.S. On the second kind twisted q-Euler numbers and polynomials of higher order. *J. Comput. Anal. Appl.* **2020**, *28*, 679–684.
8. Young, P.T. Degenerate Bernoulli polynomials, generalized factorial sums, and their applications. *J. Number Theory* **2008**, *128*, 738–758. [CrossRef]
9. Ryoo, C.S. (p,q)-analogue of Euler zeta function. *J. Appl. Math. Inf.* **2017**, *35*, 113–120. [CrossRef]
10. Ryoo, C.S. Some symmetric identities for (p,q)-Euler zeta function. *J. Comput. Anal. Appl.* **2019**, *27*, 361–366.
11. Ryoo, C.S. Some properties of the (h,p,q)-Euler numbers and polynomials and computation of their zeros. *J. Appl. Pure Math.* **2019**, *1*, 1–10.
12. Ryoo, C.S. Symmetric identities for the second kind q-Bernoulli polynomials. *J. Comput. Anal. Appl.* **2020**, *28*, 654–659.
13. Ryoo, C.S. Symmetric identities for Dirichlet-type multiple twisted (h,q)-l-function and higher-order generalized twisted (h,q)-Euler polynomials. *J. Comput. Anal. Appl.* **2020**, *28*, 537–542.
14. Riordan, J. *An Introduction to Combinatorial Analysis*; Wiley: New York, NY, USA, 1958.
15. Carlitz, L. A Note on the Multiplication Formulas for the Bernoulli and Euler Polynomials. *Proc. Am. Math. Soc.* **1953**, *4*, 184–188. [CrossRef]

© 2019 by the authors. Licensee MDPI, Basel, Switzerland. This article is an open access article distributed under the terms and conditions of the Creative Commons Attribution (CC BY) license (http://creativecommons.org/licenses/by/4.0/).

Article

Symmetric Identities for Carlitz-Type Higher-Order Degenerate (p,q)-Euler Numbers and Polynomials

Kyung-Won Hwang [1] and Cheon Seoung Ryoo [2,*]

[1] Department of Mathematics, Dong-A University, Busan 49315, Korea; khwang@dau.ac.kr
[2] Department of Mathematics, Hannam University, Daejeon 34430, Korea
* Correspondence: ryoocs@hnu.kr

Received: 18 October 2019; Accepted: 18 November 2019; Published: 20 November 2019

Abstract: The main goal of this paper is to investigate some interesting symmetric identities for Carlitz-type higher-order degenerate (p,q)-Euler numbers, and polynomials. At first, the Carlitz-type higher-order degenerate (p,q)-Euler numbers and polynomials are defined. We give few new symmetric identities for Carlitz-type higher-order degenerate (p,q)-Euler numbers and polynomials.

Keywords: Euler numbers and polynomials; degenerate Euler numbers and polynomials; Carlitz-type degenerate (p,q)-Euler numbers and polynomials; Carlitz-type higher-order degenerate (p,q)-Euler numbers and polynomials; symmetric identities

MSC: 11B68; 11S40; 11S80

---

1. Introduction

Many (p,q)-extensions of some special functions such as the hypergeometric functions, the gamma and beta functions, special polynomials, the zeta and related functions, q-series, and series representations have been studied (see [1–6]). In our paper, we always make use of the following notations: $\mathbb{Z}_+ = \mathbb{N} \cup \{0\}$ is the set of nonnegative integers, and the notation

$$\sum_{m_1,\cdots,m_r=0}^{\infty} \text{ is used instead of } \sum_{m_1=0}^{\infty} \cdots \sum_{m_r=0}^{\infty}.$$

The (p,q)-number is defined as

$$[n]_{p,q} = \frac{p^n - q^n}{p - q} = p^{n-1} + p^{n-2}q + p^{n-3}q^2 + \cdots + p^2 q^{n-3} + pq^{n-2} + q^{n-1}.$$

Much research has been conducted in the area of special functions by using (p,q)-number (see [1–6]). The classical Stirling numbers of the first kind $S_1(n,k)$ and the second kind $S_2(n,k)$ are related to each other like this (see [7–10])

$$(x)_n = \sum_{k=0}^{n} S_1(n,k)x^k \text{ and } x^n = \sum_{k=0}^{n} S_2(n,k)(x)_k,$$

respectively, where $(x)_n = x(x-1)\cdots(x-n+1)$. The generalized (p,q)-falling factorial $([x]_{p,q}|\lambda)_n$ with increment λ is defined by

$$([x]_{p,q}|\lambda)_n = \prod_{k=0}^{n-1}([x]_{p,q} - \lambda k)$$

for positive integer n, with the convention $([x]_{p,q}|\lambda)_0 = 1$; we also write

$$([x]_{p,q}|\lambda)_n = \sum_{k=0}^{n} S_1(n,k)\lambda^{n-k}[x]_{p,q}^k.$$

Clearly, $([x]_{p,q}|0)_n = [x]_{p,q}^n$. We also have the binomial theorem: for a variable x,

$$(1+\lambda t)^{\frac{[x]_{p,q}}{\lambda}} = \sum_{n=0}^{\infty} ([x]_{p,q}|\lambda)_n \frac{t^n}{n!}.$$

We introduced Carlitz-type degenerate Euler numbers $\mathcal{E}_n(\lambda)$ and Euler polynomials $\mathcal{E}_n(x,\lambda)$ using (p,q)-number (see [4]). For $0 < q < p \le 1$, $\mathcal{E}_{n,p,q}(\lambda)$ and polynomials $\mathcal{E}_{n,p,q}(x,\lambda)$ are defined by the generating functions

$$\sum_{n=0}^{\infty} \mathcal{E}_{n,p,q}(\lambda)\frac{t^n}{n!} = [2]_q \sum_{m=0}^{\infty} (-1)^m q^m (1+\lambda t)^{\frac{[m]_{p,q}}{\lambda}},$$

and

$$\sum_{n=0}^{\infty} \mathcal{E}_{n,p,q}(x,\lambda)\frac{t^n}{n!} = [2]_q \sum_{m=0}^{\infty} (-1)^m q^m (1+\lambda t)^{\frac{[m+x]_{p,q}}{\lambda}},$$

respectively (see [4]).

Hwang and Ryoo [11] discussed some properties for Carlitz-type higher-order (p,q)-Euler numbers and polynomials. For $r \in \mathbb{N}$ and $0 < q < p \le 1$, the Carlitz-type higher-order (p,q)-Euler polynomials $E_{n,p,q}^{(r)}(x)$ are defined by the generating function:

$$\sum_{n=0}^{\infty} E_{n,p,q}^{(r)}(x)\frac{t^n}{n!} = [2]_q^r \sum_{m_1,\cdots,m_r=0}^{\infty} (-1)^{m_1+\cdots+m_r} q^{m_1+\cdots+m_r} e^{[m_1+\cdots+m_r+x]_{p,q}t}. \tag{1}$$

When $x = 0$, $E_{n,p,q}^{(r)} = E_{n,p,q}^{(r)}(0)$ are called the Carlitz-type higher-order (p,q)-Euler numbers $E_{n,p,q}^{(r)}$ (see [11]). Furthermore, we obtain

$$E_{n,p,q}^{(r)}(x) = [2]_q^r \sum_{m_1,\cdots,m_r=0}^{\infty} (-1)^{m_1+\cdots+m_r} q^{m_1+\cdots+m_r} [m_1 + \cdots + m_r + x]_{p,q}^n. \tag{2}$$

For $0 < q < p \le 1$, $h \in \mathbb{Z}$, and $r \in \mathbb{N}$, Carlitz-type higher-order (h,p,q)-Euler polynomials $E_{n,p,q}^{(r,h)}(x)$ are defined using generating function

$$\sum_{n=0}^{\infty} E_{n,p,q}^{(r,h)}(x)\frac{t^n}{n!} = [2]_q^r \sum_{k_1,\cdots,k_r=0}^{\infty} (-1)^{k_1+\cdots+k_r} q^{k_1+\cdots+k_r} p^{h(k_1+\cdots+k_r)} e^{[k_1+\cdots+k_r+x]_{p,q}t}.$$

When $x = 0$, $E_{n,p,q}^{(r,h)} = E_{n,p,q}^{(r,h)}(0)$ are called the Carlitz-type higher-order (h,p,q)-Euler numbers $E_{n,p,q}^{(r,h)}$.

The following diagram shows the variations of the different types of degenerate Euler polynomials and Euler polynomials. Those polynomials in the first row and the third row of the diagram are studied by Hwang and Ryoo [4,11], Carlitz [7], Cenkci and Howard [9], Wu and Pan [12], Luo [13], and Srivastava [14], respectively. The study of these has produced beneficial results in combinatorics and number theory (see [4,7,9,12–18]). The motivation of this paper is to investigate some explicit identities and symmetric identities for Carlitz-type higher-order degenerate (p,q)-Euler polynomials in the second row of the diagram.

$$\sum_{n=0}^{\infty} \mathcal{E}_n(x,\lambda)\frac{t^n}{n!} = \left(\frac{2}{(1+\lambda t)^{\frac{1}{\lambda}}+1}\right)(1+\lambda t)^{\frac{x}{\lambda}}$$

(degenerate Euler polynomials)

$$\sum_{n=0}^{\infty} \mathcal{E}_{n,p,q}(x,\lambda)\frac{t^n}{n!} = [2]_q \sum_{m=0}^{\infty}(-1)^m q^m (1+\lambda t)^{\frac{[m+x]_{p,q}}{\lambda}}$$

(Carlitz-type degenerate (p,q)-Euler polynomials)

$$\sum_{n=0}^{\infty} \mathcal{E}_n^{(r)}(x,\lambda)\frac{t^n}{n!} = \left(\frac{2}{(1+\lambda t)^{\frac{1}{\lambda}}+1}\right)^r (1+\lambda t)^{\frac{x}{\lambda}}$$

(higher-order degenerate Euler polynomials)

$$\sum_{n=0}^{\infty} \mathcal{E}_{n,p,q}^{(r)}(x,\lambda)\frac{t^n}{n!} = [2]_q^r \sum_{m_1,\cdots,m_r=0}^{\infty}(-1)^{m_1+\cdots+m_r} q^{m_1+\cdots+m_r}(1+\lambda t)^{\frac{[m_1+\cdots+m_r+x]_{p,q}}{\lambda}}$$

(Carlitz-type higher-order degenerate (p,q)-Euler polynomials)

$$\sum_{n=0}^{\infty} E_n^{(r)}(x)\frac{t^n}{n!} = \left(\frac{2}{e^t+1}\right)^r e^{xt}$$

(higher-order Euler polynomials)

$$\sum_{n=0}^{\infty} E_{n,p,q}^{(r)}(x)\frac{t^n}{n!} = [2]_q^r \sum_{m_1,\cdots,m_r=0}^{\infty}(-1)^{m_1+\cdots+m_r} q^{m_1+\cdots+m_r}e^{[m_1+\cdots+m_r+x]_{p,q}t}$$

(Carlitz-type higher-order (p,q)-Euler polynomials)

The goal of this paper is that new generalizations of the Carlitz-type degenerate (p,q)-Euler numbers and polynomials is introduced and studied. Each section has the following contents. In Section 2, Carlitz-type higher-order degenerate (p,q)-Euler numbers and polynomials are defined. We induce some of their properties involved distribution relation, explicit formula, and so on. In Section 3, we make several symmetric identities about Carlitz-type higher-order degenerate (p,q)-Euler numbers and polynomials.

2. Carlitz-Type Higher-Order Degenerate (p,q)-Euler Numbers and Polynomials

At first, the Carlitz-type higher-order degenerate (p,q)-Euler numbers and polynomials are defined like this:

Definition 1. *For positive integer n and $r \in \mathbb{N}$, the classical higher-order Euler numbers $\mathcal{E}_n^{(r)}(\lambda)$ and Euler polynomials $\mathcal{E}_n^{(r)}(x,\lambda)$ are defined by using generating functions*

$$\left(\frac{2}{(1+\lambda t)^{\frac{1}{\lambda}}+1}\right)^r = \sum_{n=0}^{\infty} \mathcal{E}_n^{(r)}(\lambda)\frac{t^n}{n!},$$

and

$$\left(\frac{2}{(1+\lambda t)^{\frac{1}{\lambda}}+1}\right)^r (1+\lambda t)^{\frac{x}{\lambda}} = \sum_{n=0}^{\infty} \mathcal{E}_n^{(r)}(x,\lambda)\frac{t^n}{n!},$$

respectively (see [9,12]).

Now, new generalizations of the Carlitz-type degenerate (p,q)-Euler numbers and polynomials are introduced. As we have done so far, the Carlitz-type higher-order (p,q)-Euler polynomials can be defined as:

Definition 2. *For $r \in \mathbb{N}$, the Carlitz-type higher-order degenerate (p,q)-Euler numbers $\mathcal{E}_{n,p,q}^{(r)}(\lambda)$ and polynomials $\mathcal{E}_{n,p,q}^{(r)}(x,\lambda)$ are defined by using generating functions, where $0 < q < p \le 1$.*

$$\sum_{n=0}^{\infty} \mathcal{E}_{n,p,q}^{(r)}(\lambda)\frac{t^n}{n!} = [2]_q^r \sum_{m_1,\cdots,m_r=0}^{\infty} (-1)^{m_1+\cdots+m_r} q^{m_1+\cdots+m_r}(1+\lambda t)^{\dfrac{[m_1+\cdots+m_r]_{p,q}}{\lambda}},$$

and

$$\sum_{n=0}^{\infty} \mathcal{E}_{n,p,q}^{(r)}(x,\lambda)\frac{t^n}{n!} = [2]_q^r \sum_{m_1,\cdots,m_r=0}^{\infty} (-1)^{m_1+\cdots+m_r} q^{m_1+\cdots+m_r}(1+\lambda t)^{\dfrac{[m_1+\cdots+m_r+x]_{p,q}}{\lambda}},$$

respectively.

Observe that, if $p = 1, q \to 1$, then $\mathcal{E}_{n,p,q}^{(r)}(\lambda) \to \mathcal{E}_n^{(r)}(\lambda)$ and $\mathcal{E}_{n,p,q}^{(r)}(x,\lambda) \to \mathcal{E}_n^{(r)}(x,\lambda)$. Note that, if $r = 1$, then $\mathcal{E}_{n,p,q}^{(r)}(\lambda) = \mathcal{E}_{n,p,q}(\lambda)$ and $\mathcal{E}_{n,p,q}^{(r)}(x) = \mathcal{E}_{n,p,q}(x)$. If $\lambda = 0$, we have the Carlitz-type higher-order (p,q)-Euler polynomials $E_{n,p,q}^{(r)}(x)$.

By binomial theorem, we note that

$$(1+\lambda t)^{\dfrac{[m_1+\cdots+m_r+x]_{p,q}}{\lambda}}$$
$$= \sum_{k=0}^{\infty} \binom{\dfrac{[m_1+\cdots+m_r+x]_{p,q}}{\lambda}}{k} \lambda^k t^k$$
$$= \sum_{k=0}^{\infty} \left(\frac{1}{\lambda}[m_1+\cdots+m_r+x]_{p,q}\right)_k \lambda^k \frac{t^k}{k!}$$
$$= \sum_{k=0}^{\infty} \left(\frac{1}{\lambda}[m_1+\cdots+m_r+x]_{p,q}\right)\left(\frac{1}{\lambda}[m_1+\cdots+m_r+x]_{p,q}-1\right)$$
$$\cdots\left(\frac{1}{\lambda}[m_1+\cdots+m_r+x]_{p,q}-(k-1)\right)\lambda^k \frac{t^k}{k!}$$
$$= \sum_{k=0}^{\infty} \left([m_1+\cdots+m_r+x]_{p,q}\right)\left([m_1+\cdots+m_r+x]_{p,q}-\lambda\right)$$
$$\cdots\left([m_1+\cdots+m_r+x]_{p,q}-(k-1)\lambda\right)\frac{t^k}{k!}$$
$$= \sum_{k=0}^{\infty} \left([m_1+\cdots+m_r+x]_{p,q}|\lambda\right)_k \frac{t^k}{k!},$$
\hfill (3)

where generalized (p,q)-falling factorial $([x]_{p,q}|\lambda)_k = [x]_{p,q}([x]_{p,q}-\lambda)\cdots([x]_{p,q}-(k-1)\lambda)$. By Definition 2, we have the theorem below.

Theorem 1. *If $r \in \mathbb{N}$, we have*

$$\mathcal{E}_{n,p,q}^{(r)}(x,\lambda) = [2]_q^r \sum_{m_1,\cdots,m_r=0}^{\infty} (-1)^{m_1+\cdots+m_r} q^{m_1+\cdots+m_r}\left([m_1+\cdots+m_r+x]_{p,q}|\lambda\right)_n.$$

Proof. By (3), we have

$$\sum_{l=0}^{\infty} \mathcal{E}_{l,p,q}^{(r)}(x,\lambda)\frac{t^l}{l!} = [2]_q^r \sum_{m_1,\cdots,m_r=0}^{\infty} (-1)^{m_1+\cdots+m_r} q^{m_1+\cdots+m_r}(1+\lambda t)^{\dfrac{[m_1+\cdots+m_r+x]_{p,q}}{\lambda}}$$
$$= \sum_{l=0}^{\infty} \left([2]_q^r \sum_{m_1,\cdots,m_r=0}^{\infty} (-1)^{m_1+\cdots+m_r} q^{m_1+\cdots+m_r}\left([m_1+\cdots+m_r+x]_{p,q}|\lambda\right)_l\right)\frac{t^l}{l!}.$$

The first part of the theorem follows when we compare the coefficients of $\frac{t^l}{l!}$ in the above equation. We prove Theorem 1. $\square$

Note that

$$\left([m_1 + \cdots + m_r + x]_{p,q}|\lambda\right)_n = \sum_{l=0}^{n} S_1(n,l)\lambda^{n-l}[m_1 + \cdots + m_r + x]_{p,q}^l, \tag{4}$$

where $S_1(n,l)$ is the Stirling numbers of the first kind.

The relation between Carlitz-type high order degenerate (p,q)-Euler polynomials $\mathcal{E}_{n,p,q}^{(r)}(x,\lambda)$ and Carlitz-type high order (p,q)-Euler polynomials $E_{n,p,q}^{(r)}(x)$ is given by the below theorem.

Theorem 2. *For $r \in \mathbb{N}$ and $n \in \mathbb{Z}_+$, we have*

$$\mathcal{E}_{n,p,q}^{(r)}(x,\lambda) = \sum_{l=0}^{n} S_1(n,l)\lambda^{n-l}E_{l,p,q}^{(r)}(x), \quad \mathcal{E}_{n,p,q}^{(r)}(\lambda) = \sum_{l=0}^{n} S_1(n,l)\lambda^{n-l}E_{l,p,q}^{(r)}.$$

Proof. By Theorem 1, (2), and (4), we get

$$\mathcal{E}_{n,p,q}^{(r)}(x,\lambda) = \sum_{l=0}^{n} S_1(n,l)\lambda^{n-l}[2]_q^r \sum_{m_1,\cdots,m_r=0}^{\infty} (-1)^{m_1+\cdots+m_r} q^{m_1+\cdots+m_r}[m_1 + \cdots + m_r + x]_{p,q}^l$$

$$= \sum_{l=0}^{n} S_1(n,l)\lambda^{n-l}E_{l,p,q}^{(r)}(x).$$

One can obtain the desired result immediately. $\square$

The Carlitz-type higher-order degenerate (p,q)-Euler number $\mathcal{E}_{n,p,q}(\lambda)$ can be determined explicitly. A few of them are

$$\mathcal{E}_{0,p,q}^{(r)}(\lambda) = 1,$$

$$\mathcal{E}_{1,p,q}^{(r)}(\lambda) = \frac{[2]_q^r}{p-q}\left(\frac{1}{1+pq}\right)^r - \frac{[2]_q}{p-q}\left(\frac{1}{1+q^2}\right)^r,$$

$$\mathcal{E}_{2,p,q}^{(r)}(\lambda) = -\frac{[2]_q^r\lambda}{p-q}\left(\frac{1}{1+pq}\right)^r + \frac{[2]_q}{(p-q)^2}\left(\frac{1}{1+p^2q}\right)^r + \frac{[2]_q^r\lambda}{p-q}\left(\frac{1}{1+q^2}\right)^r$$
$$- \frac{2[2]_q^r}{(p-q)^2}\left(\frac{1}{1+pq^2}\right)^r + \frac{[2]_q^r}{(p-q)^2}\left(\frac{1}{1+q^3}\right)^r,$$

$$\mathcal{E}_{3,p,q}^{(r)}(\lambda) = \frac{2[2]_q^r\lambda^2}{(p-q)}\left(\frac{1}{1+pq}\right)^r - \frac{3[2]_q^r\lambda}{(p-q)^2}\left(\frac{1}{1+p^2q}\right)^r + \frac{[2]_q^r}{(p-q)^3}\left(\frac{1}{1+p^3q}\right)^r$$
$$- \frac{2[2]_q^r\lambda^2}{(p-q)}\left(\frac{1}{1+q^2}\right)^r + \frac{6[2]_q^r\lambda}{(p-q)^2}\left(\frac{1}{1+pq^2}\right)^r - \frac{3[2]_q^r}{(p-q)^3}\left(\frac{1}{1+p^2q^2}\right)^r$$
$$- \frac{3[2]_q^r\lambda}{(p-q)^2}\left(\frac{1}{1+q^3}\right)^r + \frac{3[2]_q^r}{(p-q)^3}\left(\frac{1}{1+pq^3}\right)^r - \frac{[2]_q^r}{(p-q)^3}\left(\frac{1}{1+q^4}\right)^r.$$

By using computer, Carlitz-type higher-order degenerate (p,q)-Euler number $\mathcal{E}_{n,p,q}^{(r)}(\lambda)$ can be determined explicitly. The first few $\mathcal{E}_{n,p,q}^{(r)}(\lambda)$ and $E_{n,p,q}^{(r)}$ are listed in Table 1.

Table 1. The first few numbers $\mathcal{E}^{(r)}_{n,p,q}(\lambda)$ and $E^{(r)}_{n,p,q}$.

Degree n	$\mathcal{E}^{(2)}_{n,1/2,1/3}\left(\dfrac{9}{10}\right)$	$\mathcal{E}^{(2)}_{n,1/2,1/3}\left(\dfrac{1}{10}\right)$	$\mathcal{E}^{(2)}_{n,1/2,1/3}\left(\dfrac{1}{100}\right)$	$E^{(2)}_{n,1/2,1/3}$
1	$-\dfrac{984}{1225}$	$-\dfrac{984}{1225}$	$-\dfrac{984}{1225}$	$-\dfrac{984}{1225}$
2	$-\dfrac{6149664}{53382875}$	$-\dfrac{283179072}{373680125}$	$-\dfrac{1550969286}{1868400625}$	$-\dfrac{2505564}{2989441}$
3	$\dfrac{43455323971646694}{520267306514580625}$	$-\dfrac{334418269722928746}{520267306514580625}$	$-\dfrac{11096966497657123158}{13006682662864515625}$	$-\dfrac{152830161504}{174034980625}$

Note that the limit of $\mathcal{E}^{(2)}_{n,1/2,1/3}(\lambda)$ is $E^{(2)}_{n,1/2,1/3}$ as λ approaches 0 (see Table 1).

Again, we give a relation between Carlitz-type higher-order (p,q)-Euler polynomials $E^{(r)}_{n,p,q}(x)$ and Carlitz-type higher-order degenerate (p,q)-Euler polynomials $\mathcal{E}^{(r)}_{n,p,q}(x,\lambda)$ in the theorem below.

Theorem 3. *For $m \in \mathbb{Z}_+$, we have*

$$E^{(r)}_{m,p,q}(x) = \sum_{n=0}^{m} \mathcal{E}^{(r)}_{n,p,q}(x,\lambda)\lambda^{m-n}S_2(m,n).$$

Proof. We use t instead of $\dfrac{e^{\lambda t}-1}{\lambda}$ in Definition 2, we have

$$\sum_{m=0}^{\infty} E^{(r)}_{m,p,q}(x)\frac{t^m}{m!} = [2]^r_q \sum_{m_1,\cdots,m_r=0}^{\infty} (-1)^{m_1+\cdots+m_r} q^{m_1+\cdots+m_r} e^{[m_1+\cdots+m_r+x]_{p,q}t}$$

$$= \sum_{n=0}^{\infty} \mathcal{E}^{(r)}_{n,p,q}(x,\lambda)\left(\frac{e^{\lambda t}-1}{\lambda}\right)^n \frac{1}{n!}$$

$$= \sum_{n=0}^{\infty} \mathcal{E}^{(r)}_{n,p,q}(x,\lambda)\lambda^{-n} \sum_{m=n}^{\infty} S_2(m,n)\lambda^m \frac{t^m}{m!}$$

$$= \sum_{m=0}^{\infty} \left(\sum_{n=0}^{m} \mathcal{E}^{(r)}_{n,p,q}(x,\lambda)\lambda^{m-n}S_2(m,n)\right)\frac{t^m}{m!}.$$

Use t instead of $\log(1+\lambda t)^{1/\lambda}$ in (1), we have

$$\sum_{n=0}^{\infty} E^{(r)}_{n,p,q}(x)\left(\log(1+\lambda t)^{1/\lambda}\right)^n \frac{1}{n!}$$

$$= [2]^r_q \sum_{m_1,\cdots,m_r=0}^{\infty} (-1)^{m_1+\cdots+m_r} q^{m_1+\cdots+m_r} (1+\lambda t)^{\dfrac{[m_1+\cdots+m_r+x]_{p,q}}{\lambda}} \tag{5}$$

$$= \sum_{m=0}^{\infty} \mathcal{E}^{(r)}_{m,p,q}(x,\lambda)\frac{t^m}{m!},$$

and

$$\sum_{n=0}^{\infty} E^{(r)}_{n,p,q}(x)\left(\log(1+\lambda t)^{1/\lambda}\right)^n \frac{1}{n!} = \sum_{m=0}^{\infty} \left(\sum_{n=0}^{m} E^{(r)}_{n,p,q}(x)\lambda^{m-n}S_1(m,n)\right)\frac{t^m}{m!}. \tag{6}$$

Thus, we have the theorem below from (5) and (6). $\square$

Theorem 4. *For $m \in \mathbb{Z}_+$, we have*

$$\mathcal{E}_{m,p,q}^{(r)}(x,\lambda) = \sum_{n=0}^{m} E_{n,p,q}^{(r)}(x)\lambda^{m-n}S_1(m,n).$$

We note that

$$(1+\lambda t)^{\dfrac{[m_1+\cdots+m_r+x]_{p,q}}{\lambda}}$$

$$= (1+\lambda t)^{\dfrac{p^x[m_1+\cdots+m_r]_{p,q}}{\lambda}}(1+\lambda t)^{\dfrac{q^{m_1+\cdots+m_r}[x]_{p,q}}{\lambda}}$$

$$= \sum_{m=0}^{\infty}(p^x[m_1+\cdots+m_r]_{p,q}|\lambda)_m\frac{t^m}{m!}e^{\log(1+\lambda t)^{\dfrac{q^{m_1+\cdots+m_r}[x]_{p,q}}{\lambda}}}$$

$$= \sum_{m=0}^{\infty}(p^x[m_1+\cdots+m_r]_{p,q}|\lambda)_m\frac{t^m}{m!}\sum_{l=0}^{\infty}\left(\frac{q^{m_1+\cdots+m_r}[x]_{p,q}}{\lambda}\right)^l\frac{\log(1+\lambda t)^l}{l!}$$

$$= \sum_{m=0}^{\infty}(p^x[m_1+\cdots+m_r]_{p,q}|\lambda)_m\frac{t^m}{m!}\sum_{l=0}^{\infty}\left(\frac{q^{m_1+\cdots+m_r}[x]_{p,q}}{\lambda}\right)^l\sum_{k=l}^{\infty}S_1(k,l)\lambda^k\frac{t^k}{k!}$$

$$= \sum_{n=0}^{\infty}\left(\sum_{k=0}^{n}\sum_{l=0}^{k}\binom{n}{k}(p^x[m_1+\cdots+m_r]_{p,q}|\lambda)_{n-k}\lambda^{k-l}q^{(m_1+\cdots+m_r)l}[x]_{p,q}^l S_1(k,l)\right)\frac{t^n}{n!}.$$

$$(7)$$

By Definition 2 and (7), we get

$$\sum_{n=0}^{\infty}\mathcal{E}_{n,p,q\zeta}^{(r)}(x,\lambda)\frac{t^n}{n!}$$

$$= [2]_q^r\sum_{m_1,\cdots,m_r=0}^{\infty}(-1)^{m_1+\cdots+m_r}q^{m_1+\cdots+m_r}(1+\lambda t)^{\dfrac{[m_1+\cdots+m_r+x]_{p,q}}{\lambda}}$$

$$= [2]_q^r\sum_{m_1,\cdots,m_r=0}^{\infty}(-1)^{m_1+\cdots+m_r}q^{m_1+\cdots+m_r}$$

$$\times \sum_{n=0}^{\infty}\left(\sum_{k=0}^{n}\sum_{l=0}^{k}\binom{n}{k}(p^x[m_1+\cdots+m_r]_{p,q}|\lambda)_{n-k}\lambda^{k-l}q^{(m_1+\cdots+m_r)l}[x]_{p,q}^l S_1(k,l)\right)\frac{t^n}{n!}.$$

When we compare the coefficients of $\frac{t^n}{n!}$ in the above equation, we have the theorem below.

Theorem 5. *For $0 < q < p \leq 1$, $r \in \mathbb{N}$, and $n \in \mathbb{Z}_+$,*

$$\mathcal{E}_{n,p,q}^{(r)}(x,\lambda) = [2]_q^r\sum_{m_1,\cdots,m_r=0}^{\infty}\sum_{k=0}^{n}\sum_{l=0}^{k}\binom{n}{k}(-1)^{m_1+\cdots+m_r}q^{m_1+\cdots+m_r}$$

$$\times (p^x[m_1+\cdots+m_r]_{p,q}|\lambda)_{n-k}\lambda^{k-l}q^{(m_1+\cdots+m_r)l}[x]_{p,q}^l S_1(k,l).$$

From (4) and Theorem 2, we get this:

$$\sum_{n=0}^{\infty} \mathcal{E}_{n,p,q}^{(r)}(x,\lambda)\frac{t^n}{n!}$$

$$= [2]_q^r \sum_{m_1,\cdots,m_r=0}^{\infty} (-1)^{m_1+\cdots+m_r} q^{m_1+\cdots+m_r} (1+\lambda t)^{\dfrac{[m_1+\cdots+m_r+x]_{p,q}}{\lambda}}$$

$$= [2]_q^r \sum_{m_1,\cdots,m_r=0}^{\infty} (-1)^{m_1+\cdots+m_r} q^{m_1+\cdots+m_r}$$

$$\times \sum_{n=0}^{\infty}\sum_{l=0}^{n} S_1(n,l)\lambda^{n-l}\frac{\sum_{j=0}^{l}\binom{l}{j}(-1)^j p^{x(l-j)}q^{xj}}{(p-q)^l} p^{(l-j)(m_1+\cdots+m_r)} q^{j(m_1+\cdots+m_r)}\frac{t^n}{n!}$$

$$= \sum_{n=0}^{\infty}\left([2]_q^r \sum_{l=0}^{n}\sum_{j=0}^{l} \frac{S_1(n,l)\lambda^{n-l}\binom{l}{j}(-1)^j q^{xj}p^{x(l-j)}}{(p-q)^l}\left(\frac{1}{1+q^{j+1}p^{l-j}}\right)^r\right)\frac{t^n}{n!}.$$

When we compare the coefficients $\frac{t^n}{n!}$ in the above equation, we get the theorem.

Theorem 6. *For $r \in \mathbb{N}$ and $n \in \mathbb{Z}_+$,*

$$\mathcal{E}_{n,p,q}^{(r)}(x,\lambda) = [2]_q^r \sum_{l=0}^{n}\sum_{j=0}^{l} \frac{S_1(n,l)\lambda^{n-l}\binom{l}{j}(-1)^j q^{xj}p^{x(l-j)}}{(p-q)^l}\left(\frac{1}{1+q^{j+1}p^{l-j}}\right)^r,$$

$$\mathcal{E}_{n,p,q}^{(r)}(\lambda) = [2]_q^r \sum_{l=0}^{n}\sum_{j=0}^{l} \frac{S_1(n,l)\lambda^{n-l}\binom{l}{j}(-1)^j}{(p-q)^l}\left(\frac{1}{1+q^{j+1}p^{l-j}}\right)^r.$$

The Carlitz-type high order degenerate (p,q)-Euler polynomials $\mathcal{E}_{n,p,q}(x,\lambda)$ can be determined explicitly. Here are a few of them:

$$\mathcal{E}_{0,p,q}^{(r)}(x,\lambda) = 1,$$

$$\mathcal{E}_{1,p,q}^{(r)}(x,\lambda) = \frac{[2]_q^r p^x}{p-q}\left(\frac{1}{1+pq}\right)^r - \frac{[2]_q^r q^x}{p-q}\left(\frac{1}{1+q^2}\right)^r,$$

$$\mathcal{E}_{2,p,q}^{(r)}(x,\lambda) = -\frac{[2]_q^r \lambda p^x}{p-q}\left(\frac{1}{1+pq}\right)^r + \frac{[2]_q^r p^{2x}}{(p-q)^2}\left(\frac{1}{1+p^2q}\right)^r + \frac{[2]_q^r \lambda q^x}{p-q}\left(\frac{1}{1+q^2}\right)^r$$
$$-\frac{2[2]_q^r p^x q^x}{(p-q)^2}\left(\frac{1}{1+pq^2}\right)^r + \frac{[2]_q^r q^{2x}}{(p-q)^2}\left(\frac{1}{1+q^3}\right)^r,$$

$$\mathcal{E}_{3,p,q}^{(r)}(x,\lambda) = \frac{2[2]_q^r \lambda^2 p^x}{p-q}\left(\frac{1}{1+pq}\right)^r - \frac{3[2]_q^r \lambda p^{2x}}{(p-q)^2}\left(\frac{1}{1+p^2q}\right)^r + \frac{[2]_q^r p^{3x}}{(p-q)^3}\left(\frac{1}{1+p^3q}\right)^r$$
$$-\frac{2[2]_q^r \lambda^2 q^x}{p-q}\left(\frac{1}{1+q^2}\right)^r + \frac{6[2]_q^r \lambda p^x q^x}{(p-q)^2}\left(\frac{1}{1+pq^2}\right)^r - \frac{3[2]_q^r p^{2x}q^x}{(p-q)^3}\left(\frac{1}{1+p^2q^2}\right)^r$$
$$-\frac{3[2]_q^r \lambda q^{2x}}{(p-q)^2}\left(\frac{1}{1+q^3}\right)^r + \frac{3[2]_q^r p^x q^{2x}}{(p-q)^3}\left(\frac{1}{1+pq^3}\right)^r - \frac{[2]_q^r q^{3x}}{(p-q)^3}\left(\frac{1}{1+q^4}\right)^r.$$

3. Some Symmetric Identities for Carlitz-Type Higher-Order Degenerate (p,q)-Euler Numbers and Polynomials

Let $w_1 \equiv 1 \pmod 2$, $w_2 \equiv 1 \pmod 2$ for $w_1, w_2 \in \mathbb{N}$. For $r \in \mathbb{N}$ and $n \in \mathbb{Z}_+$, we obtain certain symmetry identities for Carlitz-type higher-order degenerate (p,q)-Euler numbers and polynomials.

Theorem 7. *Let $w_1 \equiv 1 \pmod 2$, $w_2 \equiv 1 \pmod 2$ for $w_1, w_2 \in \mathbb{N}$. Then, we obtain*

$$
[w_1]_{p,q}^n [2]_{q^{w_2}}^r \sum_{j_1,\cdots,j_r=0}^{w_1-1} (-1)^{j_1+\cdots+j_r} q^{w_2(j_1+\cdots+j_r)}
$$

$$
\times \, \mathcal{E}_{n,p^{w_1}q^{w_1}}^{(r)} \left(w_2 x + \frac{w_2}{w_1}(j_1+\cdots+j_r), \frac{\lambda}{[w_1]_{p,q}} \right)
$$

$$
= [w_2]_{p,q}^n [2]_{q^{w_1}}^r \sum_{j_1,\cdots,j_r=0}^{w_2-1} (-1)^{j_1+\cdots+j_r} q^{w_1(j_1+\cdots+j_r)}
$$

$$
\times \, \mathcal{E}_{n,p^{w_2},q^{w_2}}^{(r)} \left(w_1 x + \frac{w_1}{w_2}(j_1+\cdots+j_r), \frac{\lambda}{[w_2]_{p,q}} \right). \tag{8}
$$

Proof. Note that $[xy]_{p,q} = [x]_{p^y,q^y}[y]_{p,q}$ for any $x, y \in \mathbb{C}$. In Definition 2, we induce the next result by substituting $w_1 x + \frac{w_1}{w_2}(j_1+\cdots+j_r)$ instead of x and replace q, p, and λ by q^{w_2}, p^{w_2}, and $\frac{\lambda}{[w_2]_{p,q}}$, respectively:

$$
\sum_{n=0}^{\infty} \left([w_2]_{p,q}^n [2]_{q^{w_1}}^r \sum_{j_1,\cdots,j_r=0}^{w_2-1} (-1)^{\sum_{l=1}^r j_l} q^{w_1(\sum_{l=1}^r j_l)} \mathcal{E}_{n,p^{w_2},q^{w_2}}^{(r)} \left(w_1 x + \frac{w_1}{w_2}(\sum_{l=1}^r j_l), \frac{\lambda}{[w_2]_{p,q}} \right) \right) \frac{t^n}{n!}
$$

$$
= [2]_{q^{w_1}}^r \sum_{j_1,\cdots,j_r=0}^{w_2-1} (-1)^{\sum_{l=1}^r j_l} q^{w_1(\sum_{l=1}^r j_l)} \sum_{n=0}^{\infty} \mathcal{E}_{n,p^{w_2},q^{w_2}}^{(r)} \left(w_1 x + \frac{w_1}{w_2}(\sum_{l=1}^r j_l), \frac{\lambda}{[w_2]_{p,q}} \right) \frac{([w_2]_{p,q}t)^n}{n!}
$$

$$
= [2]_{q^{w_1}}^r \sum_{j_1,\cdots,j_r=0}^{w_2-1} (-1)^{\sum_{l=1}^r j_l} q^{w_1(\sum_{l=1}^r j_l)} [2]_{q^{w_2}}^r \sum_{m_1,\cdots,m_r=0}^{\infty} (-1)^{m_1+\cdots+m_r} q^{w_2(m_1+\cdots+m_r)}
$$

$$
\times \left(1 + \frac{\lambda}{[w_2]_{p,q}}[w_2]_{p,q}t \right)^{\dfrac{\left[w_1 x + w_1 x + \frac{w_1}{w_2}(j_1+\cdots+j_r) + m_1 + \cdots + m_r \right]_{p^{w_2},q^{w_2}}}{\frac{\lambda}{[w_2]_{p,q}}}}
$$

$$
= [2]_{q^{w_1}}^r \sum_{j_1,\cdots,j_r=0}^{w_2-1} (-1)^{\sum_{l=1}^r j_l} q^{w_1(\sum_{l=1}^r j_l)} [2]_{q^{w_2}}^r \sum_{m_1,\cdots,m_r=0}^{\infty} (-1)^{m_1+\cdots+m_r} q^{w_2(m_1+\cdots+m_r)}
$$

$$
\times (1 + \lambda t)^{\dfrac{[w_1 w_2 x + w_1(j_1+\cdots+j_r) + w_2(m_1+\cdots m_r)]_{p,q}}{\lambda}}.
$$

Since there exists the unique non-negative integer n such that $m = wn + i$ with $0 \leq i \leq w - 1$ for any non-negative integer m and odd positive integer w, this can be written

$$[2]_{q^{w_1}}^r [2]_{q^{w_2}}^r \sum_{j_1,\cdots,j_r=0}^{w_2-1} (-1)^{\sum_{l=1}^r j_l} q^{w_1(\sum_{l=1}^r j_l)} \sum_{m_1,\cdots,m_r=0}^{\infty} (-1)^{m_1+\cdots+m_r} q^{w_2(m_1+\cdots+m_r)}$$

$$\times (1+\lambda t)^{\dfrac{[w_1 w_2 x + w_1(j_1+\cdots+j_r) + w_2(m_1+\cdots m_r)]_{p,q}}{\lambda}}$$

$$= [2]_{q^{w_1}}^r [2]_{q^{w_2}}^r \sum_{j_1,\cdots,j_r=0}^{w_2-1} (-1)^{\sum_{l=1}^r j_l} q^{w_1(\sum_{l=1}^r j_l)}$$

$$\times \sum_{\substack{w_1 n_1+i_1,\cdots,w_1 n_r+i_r=0 \\ 0 \leq i_k \leq w_1-1 \\ 1 \leq k \leq r}}^{\infty} (-1)^{w_1 n_1+i_1+\cdots+w_1 n_r+i_r} q^{w_2(w_1 n_1+i_1+\cdots+w_1 n_r+i_r)}$$

$$\times (1+\lambda t)^{\dfrac{[w_1 w_2 x + w_1(j_1+\cdots+j_r) + w_2 w_1(n_1+\cdots+n_r) + w_2(i_1+\cdots+i_r)]_{p,q}}{\lambda}}$$

$$= [2]_{q^{w_1}}^r [2]_{q^{w_2}}^r \sum_{j_1,\cdots,j_r=0}^{w_2-1} (-1)^{\sum_{l=1}^r j_l} q^{w_1(\sum_{l=1}^r j_l)}$$

$$\times \sum_{i_1,\cdots,i_r=0}^{w_1-1} \sum_{n_1,\cdots,n_r=0}^{\infty} (-1)^{\sum_{l=1}^r n_l} (-1)^{\sum_{l=1}^r i_l} q^{w_2(\sum_{l=1}^r i_l)} q^{w_1 w_2(\sum_{l=1}^r n_l)}$$

$$\times (1+\lambda t)^{\dfrac{[w_1 w_2 x + w_1(j_1+\cdots+j_r) + w_2 w_1(n_1+\cdots+n_r) + w_2(i_1+\cdots+i_r)]_{p,q}}{\lambda}} .$$

We obtain the following formula using the formula above:

$$\sum_{n=0}^{\infty} \left([w_2]_{p,q}^n [2]_{q^{w_1}}^r \sum_{j_1,\cdots,j_r=0}^{w_2-1} (-1)^{\sum_{l=1}^r j_l} q^{w_1(\sum_{l=1}^r j_l)} \mathcal{E}_{n,p^{w_2},q^{w_2}}^{(r)} \left(w_1 x + \frac{w_1}{w_2}(\sum_{l=1}^r j_l), \frac{\lambda}{[w_2]_{p,q}} \right) \right) \frac{t^n}{n!}$$

$$= [2]_{q^{w_1}}^r [2]_{q^{w_2}}^r \sum_{n_1,\cdots,n_r=0}^{\infty} \sum_{j_1,\cdots,j_r=0}^{w_2-1} \sum_{i_1,\cdots,i_r=0}^{w_1-1} (-1)^{\sum_{l=1}^r j_l} (-1)^{\sum_{l=1}^r n_l} (-1)^{\sum_{l=1}^r i_l}$$

$$\times q^{w_1(\sum_{l=1}^r j_l)} q^{w_2(\sum_{l=1}^r i_l)} q^{w_1 w_2(\sum_{l=1}^r n_l)}$$

$$\times (1+\lambda t)^{\dfrac{[w_1 w_2 x + w_1(j_1+\cdots+j_r) + w_2 w_1(n_1+\cdots+n_r) + w_2(i_1+\cdots+i_r)]_{p,q}}{\lambda}} . \tag{9}$$

From a similar approach, we also have that

$$\sum_{n=0}^{\infty}\left([w_1]_{p,q}^n[2]_{q^{w_2}}^r\sum_{j_1,\cdots,j_r=0}^{w_1-1}(-1)^{\sum_{l=1}^{r}j_l}q^{w_2(\sum_{l=1}^{r}j_l)}\mathcal{E}_{n,p^{w_1},q^{w_1}}^{(r)}\left(w_2x+\frac{w_2}{w_1}\left(\sum_{l=1}^{r}j_l\right),\frac{\lambda}{[w_1]_{p,q}}\right)\right)\frac{t^n}{n!}$$

$$= [2]_{q^{w_2}}^r[2]_{q^{w_1}}^r\sum_{n_1,\cdots,n_r=0}^{\infty}\sum_{j_1,\cdots,j_r=0}^{w_1-1}\sum_{i_1,\cdots,i_r=0}^{w_2-1}(-1)^{\sum_{l=1}^{r}j_l}(-1)^{\sum_{l=1}^{r}n_l}(-1)^{\sum_{l=1}^{r}i_l}$$

$$\times q^{w_2(\sum_{l=1}^{r}j_l)}q^{w_1(\sum_{l=1}^{r}i_l)}q^{w_1w_2(\sum_{l=1}^{r}n_l)}$$

$$(10)$$

$$\times (1+\lambda t)^{\dfrac{[w_1w_2x+w_2(j_1+\cdots+j_r)+w_2w_1(n_1+\cdots+n_r)+w_1(i_1+\cdots+i_r)]_{p,q}}{\lambda}}.$$

Therefore, by (9) and (10), we can obtain the desired result. $\square$

Taking $w_2=1$ in Theorem 7, we obtain the following multiplication theorem for Carlitz-type higher-order degenerate (p,q)-Euler polynomials.

Theorem 8. *Let $w_1\equiv 1\pmod 2$ for $w_1\in\mathbb{N}$. For $r\in\mathbb{N}$ and $n\in\mathbb{Z}_+$, we obtain*

$$\mathcal{E}_{n,p,q}^{(r)}(w_1x,\lambda)=\frac{[2]_q^r}{[2]_{q^{w_1}}^r}[w_1]_{p,q}^n\sum_{j_1,\cdots,j_r=0}^{w_1-1}(-1)^{j_1+\cdots+j_r}q^{j_1+\cdots+j_r}$$

$$(11)$$

$$\times\mathcal{E}_{n,p^{w_1},q^{w_1}}^{(r)}\left(x+\frac{j_1+\cdots+j_r}{w_1},\frac{\lambda}{[w_1]_{p,q}}\right).$$

Taking $\lambda=0$ in (11), we get the multiplication theorem for Carlitz-type high order (p,q)-Euler polynomials (see [11]).

Corollary 1. *Let $w_1\equiv 1\pmod 2$ for $w_1\in\mathbb{N}$. For $n\in\mathbb{Z}_+$ and $r\in\mathbb{N}$, we get*

$$E_{n,p,q}^{(r)}(w_1x)=\frac{[2]_q^r}{[2]_{q^{w_1}}^r}[w_1]_{p,q}^n\sum_{j_1,\cdots,j_r=0}^{w_1-1}(-1)^{j_1+\cdots+j_r}q^{j_1+\cdots+j_r}$$

$$\times E_{n,p^{w_1},q^{w_1}}^{(r)}\left(x+\frac{j_1+\cdots+j_r}{w_1}\right).$$

For $r=1$ in (10), we have the multiplication theorem for Carlitz-type degenerate (p,q)-Euler polynomials (see [4]).

Corollary 2. *Let $w_1\equiv 1\pmod 2$ for $w_1\in\mathbb{N}$. For $n\in\mathbb{Z}_+$,*

$$\mathcal{E}_{n,p,q}(w_1x,\lambda)=\frac{[2]_q}{[2]_{q^{w_1}}}[w_1]_{p,q}^n\sum_{j=0}^{w_1-1}(-1)^jq^j\mathcal{E}_{n,p^{w_1},q^{w_1}}\left(x+\frac{j}{w_1},\frac{\lambda}{[w_1]_{p,q}}\right).$$

If $p=1,q\to 1$ in Corollary 2, then we get the corollary.

Corollary 3. *Let $m\equiv 1\pmod 2$ for $m\in\mathbb{N}$. For $n\in\mathbb{Z}_+$,*

$$\mathcal{E}_n(x,\lambda) = m^n \sum_{j=0}^{m-1} (-1)^j q^j \mathcal{E}_n \left(\frac{x+j}{m}, \frac{\lambda}{m} \right).$$
(12)

If λ approaches to 0 in (12), this leads to the distribution relation for Euler polynomials

$$E_n(x) = m^n \sum_{j=0}^{m-1} (-1)^j E_n \left(\frac{x+i}{m} \right).$$

By Theorem 2 and Theorem 7, it follows the theorem below.

Theorem 9. *Let w_1 and w_2 be odd positive integers. Then, it has*

$$\sum_{l=0}^{n} S_1(n,l) \lambda^{n-l} [w_1]_{p,q}^l [2]_{q^{w_2}}^r$$

$$\times \sum_{j_1,\cdots,j_r=0}^{w_1-1} (-1)^{j_1+\cdots+j_r} q^{w_2(j_1+\cdots+j_r)} E_{l,p^{w_1},q^{w_1}}^{(r)} \left(w_2 x + \frac{w_2}{w_1}(j_1+\cdots+j_r) \right)$$

$$= \sum_{l=0}^{n} S_1(n,l) \lambda^{n-l} [w_2]_{p,q}^l [2]_{q^{w_1}}^r$$

$$\times \sum_{j_1,\cdots,j_r=0}^{w_2-1} (-1)^{j_1+\cdots+j_r} q^{w_1(j_1+\cdots+j_r)} E_{l,p^{w_2},q^{w_2}}^{(r)} \left(w_1 x + \frac{w_1}{w_2}(j_1+\cdots+j_r) \right).$$

We get another symmetry identity by using the addition theorem about the Carlitz-type higher-order degenerate (p,q)-Euler polynomials $\mathcal{E}_{n,p,q}^{(r)}(x)$. Let

$$A_{n,k,p,q}^{(r)}(w) = \sum_{j_1,\cdots,j_r=0}^{w-1} (-1)^{\sum_{i=1}^{r} j_i} q^{(n-k+1)(\sum_{i=1}^{r} j_i)} [j_1 \cdots + j_k]_{p,q}^k$$

for each integer $n \geq 0$. The $A_{n,k,p,q}^{(k)}(w)$ is called as the alternating (p,q)-sums of powers.

Theorem 10. *Let $w_1, w_2 \in \mathbb{N}$ with $w_1 \equiv 1 \pmod{2}$, $w_2 \equiv 1 \pmod{2}$. For $r \in \mathbb{N}$ and $n \in \mathbb{Z}_+$, we obtain*

$$\sum_{l=0}^{n} \sum_{k=0}^{l} \binom{l}{k} S_1(n,l) \lambda^{n-l} p^{w_1 w_2 x k} [2]_{q^{w_1}} [w_1]_{p,q}^k [w_2]_{p,q}^{l-k} E_{l-k,p^{w_2},q^{w_2}}^{(r,k)} (w_1 x) A_{l,k,p^{w_1},q^{w_1}}^{(r)}(w_2)$$

$$= \sum_{l=0}^{n} \sum_{k=0}^{l} \binom{l}{k} S_1(n,l) \lambda^{n-l} p^{w_1 w_2 x k} [2]_{q^{w_2}} [w_2]_{p,q}^k [w_1]_{p,q}^{l-k} E_{l-k,p^{w_1},q^{w_1}}^{(r,k)} (w_2 x) A_{l,k,p^{w_2},q^{w_2}}^{(r)}(w_1).$$

Proof. Now, we use the addition theorem about the Carlitz-type higher-order degenerate (p,q)-Euler polynomials (see [10]). We derive

$$\sum_{j_1,\cdots,j_r=0}^{w_1-1} (-1)^{\sum_{i=1}^{r} j_i} q^{w_2(\sum_{i=1}^{r} j_i)} E_{l,p^{w_1},q^{w_1}}^{(r)}\left(w_2 x + \frac{w_2}{w_1}(j_1+\cdots+j_k)\right)$$

$$= \sum_{j_1,\cdots,j_r=0}^{w_1-1} (-1)^{\sum_{i=1}^{r} j_i} q^{w_2(\sum_{i=1}^{r} j_i)}$$

$$\times \sum_{k=0}^{l} \binom{l}{k} q^{w_2(l-k)(\sum_{i=1}^{r} j_i)} p^{w_1 w_2 x k} E_{l-k,p^{w_1},q^{w_1}}^{(r,k)}(w_2 x) \left[\frac{w_2}{w_1}(j_1+\cdots+j_r)\right]_{p^{w_1},q^{w_1}}^{k}$$

$$= \sum_{j_1,\cdots,j_r=0}^{w_1-1} (-1)^{\sum_{i=1}^{r} j_i} q^{w_2(\sum_{i=1}^{r} j_i)}$$

$$\times \sum_{k=0}^{l} \binom{l}{k} q^{w_2(l-k)(\sum_{i=1}^{r} j_i)} p^{w_1 w_2 x k} E_{l-k,p^{w_1},q^{w_1}}^{(r,k)}(w_2 x) \left(\frac{[w_2]_{p,q}}{[w_1]_{p,q}}\right)^{k} [j_1+\cdots+j_r]_{p^{w_2},q^{w_2}}^{k}.$$

By Theorem 12, then we have

$$\sum_{l=0}^{n} S_1(n,l) \lambda^{n-l} [w_1]_{p,q}^{l} [2]_{q^{w_2}}^{r}$$

$$\times \sum_{j_1,\cdots,j_r=0}^{w_1-1} (-1)^{j_1+\cdots+j_r} q^{w_2(j_1+\cdots+j_r)} E_{l,p^{w_1},q^{w_1}}^{(r)}\left(w_2 x + \frac{w_2}{w_1}(j_1+\cdots+j_r)\right)$$

$$= \sum_{l=0}^{n}\sum_{k=0}^{l} \binom{l}{k} S_1(n,l) \lambda^{n-l} [w_1]_{p,q}^{l-k} [w_2]_{p,q}^{k} [2]_{q^{w_2}}^{r} p^{w_1 w_2 x k} E_{l-k,p^{w_1},q^{w_1}}^{(r,k)}(w_2 x) \qquad (13)$$

$$\times \sum_{j_1,\cdots,j_r=0}^{w_1-1} (-1)^{\sum_{i=1}^{r} j_i} q^{w_2(l-k+1)(\sum_{i=1}^{r} j_i)} [j_1+\cdots+j_r]_{p^{w_2},q^{w_2}}^{k}$$

$$= \sum_{l=0}^{n}\sum_{k=0}^{l} \binom{l}{k} S_1(n,l) \lambda^{n-l} [w_1]_{p,q}^{l-k} [w_2]_{p,q}^{k} [2]_{q^{w_2}}^{r} p^{w_1 w_2 x k} E_{l-k,p^{w_1},q^{w_1}}^{(r,k)}(w_2 x) \mathcal{A}_{l,k,p^{w_2},q^{w_2}}^{(k)}(w_2).$$

Similarly, we have

$$\sum_{l=0}^{n} S_1(n,l) \lambda^{n-l} [w_2]_{p,q}^{l} [2]_{q^{w_1}}^{r}$$

$$\times \sum_{j_1,\cdots,j_r=0}^{w_2-1} (-1)^{j_1+\cdots+j_r} q^{w_1(j_1+\cdots+j_r)} E_{l,p^{w_2},q^{w_2}}^{(r)}\left(w_1 x + \frac{w_1}{w_2}(j_1+\cdots+j_r)\right)$$

$$= \sum_{l=0}^{n}\sum_{k=0}^{l} \binom{l}{k} S_1(n,l) \lambda^{n-l} [w_2]_{p,q}^{l-k} [w_1]_{p,q}^{k} [2]_{q^{w_1}}^{r} p^{w_1 w_2 x k} E_{l-k,p^{w_2},q^{w_2}}^{(r,k)}(w_1 x) \qquad (14)$$

$$\times \sum_{j_1,\cdots,j_r=0}^{w_2-1} (-1)^{\sum_{i=1}^{r} j_i} q^{w_1(l-k+1)(\sum_{i=1}^{r} j_i)} [j_1+\cdots+j_r]_{p^{w_1},q^{w_1}}^{k}$$

$$= \sum_{l=0}^{n}\sum_{k=0}^{l} \binom{l}{k} S_1(n,l) \lambda^{n-l} [w_2]_{p,q}^{l-k} [w_1]_{p,q}^{k} [2]_{q^{w_1}}^{r} p^{w_1 w_2 x k} E_{l-k,p^{w_2},q^{w_2}}^{(r,k)}(w_1 x) \mathcal{A}_{l,k,p^{w_1},q^{w_1}}^{(k)}(w_2).$$

By (13) and (14), we make the desired symmetric identity. $\square$

By Theorem 10, we have the symmetric identity for the Carlitz-type high order (h,p,q)-Euler numbers $E_{n,p,q}^{(r,h)}$ in complex field.

Corollary 4. *Let $w_1 \equiv 1 \pmod 2$, $w_2 \equiv 1 \pmod 2$, where $w_1, w_2 \in \mathbb{N}$. For $r \in \mathbb{N}$ and $n \in \mathbb{Z}_+$, we obtain*

$$\sum_{l=0}^{n} \sum_{k=0}^{l} \binom{l}{k} S_1(n,l) \lambda^{n-l} [2]_{q^{w_1}} [w_1]_{p,q}^{k} [w_2]_{p,q}^{l-k} \mathcal{A}_{l,k,p^{w_1},q^{w_1}}^{(r)} (w_2) E_{l-k,p^{w_2},q^{w_2}}^{(r,k)}$$

$$= \sum_{l=0}^{n} \sum_{k=0}^{l} \binom{l}{k} S_1(n,l) \lambda^{n-l} [2]_{q^{w_2}} [w_2]_{p,q}^{k} [w_1]_{p,q}^{l-k} \mathcal{A}_{l,k,p^{w_2},q^{w_2}}^{(r)} (w_1) E_{l-k,p^{w_1},q^{w_1}}^{(r,k)}.$$

4. Conclusions

In our previous paper [4], we studied some identities of symmetry on the Carlitz-type degenerate (p,q)-Euler polynomials. The motivation of this paper is to investigate some explicit identities for the Carlitz-type higher-order degenerate (p,q)-Euler polynomials in the second row of the diagram at page 3. Thus, we defined the Carlitz-type higher-order degenerate (p,q)-Euler polynomials in Definition 2 and obtained the formulas (explicit formula (Theorem 6), multiplication theorem (Theorem 8), and distribution relation (Corollary 2, Corollary 3)). In Theorem 7, we gave some symmetry identities for the Carlitz-type higher-order degenerate (p,q)-Euler polynomials. We also obtained the explicit identities related to the Carlitz-type higher-order (p,q)-Euler polynomials, the alternating (p,q)-sums of powers, and Stirling numbers (see Theorem 10 and Corollary 4). In particular, these results generalized some well-known properties relating degenerate Euler numbers and polynomials, degenerate Stirling numbers, alternating sums of powers, multiplication theorem, distribution relation, falling factorial, symmetry properties of the degenerate Euler numbers and polynomials (see [7–18]). In addition, in this paper, if we take $r = 1$, then [4] is the special case of this paper.

Author Contributions: These authors contributed equally to this work.

Funding: This work was supported by the Dong-A university research fund.

Acknowledgments: The authors would like to thank the referees for their valuable comments, which improved the original manuscript in its present form.

Conflicts of Interest: The authors declare no conflicts of interest.

References

1. Agarwal, R.P.; Kang, J.Y.; Ryoo, C.S. Some properties of (p,q)-tangent polynomials. *J. Comput. Anal. Appl.* **2018**, *24*, 1439–1454.
2. Araci, S.; Duran, U.; Acikgoz, M.; Srivastava, H.M. A certain (p,q)-derivative operato rand associated divided differences. *J. Inequal. Appl.* **2016**, *2016*, 301. [CrossRef]
3. Duran, U.; Acikgoz, M.; Araci, S. On (p,q)-Bernoulli, (p,q)-Euler and (p,q)-Genocchi polynomials. *J. Comput. Theor. Nanosci.* **2016**, *13*, 7833–7846. [CrossRef]
4. Hwang, K.W.; Ryoo, C.S. Some symmetric identities for degenerate Carlitz-type (p,q)-Euler numbers and polynomials. *Symmetry* **2019**, *11*, 830, doi:10.3390/sym11060830. [CrossRef]
5. Ryoo, C.S. (p,q)-analogue of Euler zeta function. *J. Appl. Math. Inf.* **2017**, *35*, 113–120. [CrossRef]
6. Ryoo, C.S. Some symmetric identities for (p,q)-Euler zeta function. *J. Comput. Anal. Appl.* **2019**, *27*, 361–366.
7. Carlitz, L. Degenerate Stiling, Bernoulli and Eulerian numbers. *Util. Math.* **1979**, *15*, 51–88.
8. Young, P. T. Degenerate Bernoulli polynomials, generalized factorial sums, and their applications. *J. Number Theor.* **2008**, *128*, 738–758. [CrossRef]
9. Cenkci, M.; Howard, F.T. Notes on degenerate numbers. *Discret. Math.* **2007**, *307*, 2395–2375. [CrossRef]
10. Howard, F.T. Explicit formulas for degenerate Bernoulli numbers. *Discret. Math.* **1996**, *162*, 175–185. [CrossRef]
11. Hwang, K.W.; Ryoo, C.S. Some symmetric identities for the multiple (p,q)-Hurwitz-Euler eta function. *Symmetry* **2019**, *11*, 645. [CrossRef]
12. Wu, M.; Pan, H. Sums of products of the degenerate Euler numbers. *Adv. Differ. Equ.* **2014**, *40*. [CrossRef]

13. Luo, Q.M. The multiplication formulas for the Apostol-Bernoulli and Apostol-Euler polynomials of higher order. *Integral Trans. Spec. Funct.* **2009**, *20*, 377–391. [CrossRef]
14. Srivastava, H.M. Some generalizations and basic (or q-) extensions of the Bernoulli, Euler and Genocchi Polynomials. *Appl. Math. Inf. Sci.* **2011**, *5*, 390–444.
15. Andrews, L.C. *Special Functions for Engineers and Mathematicians*; Macmillan Co.: New York, NY, USA, 1985.
16. Andrews, G.E.; Askey, R.; Roy, R. Special Functions. In *Encyclopedia of Mathematics and Its Applications 71*; Cambridge University Press: Cambridge, UK, 1999.
17. Erdelyi, A.; Magnus, W.; Oberhettinger, F.; Tricomi, F.G. *Higher Transcendental Functions*; Krieger: New York, NY, USA, 1981; Volume 3.
18. Robert, A.M. A Course in p-adic Analysis. In *Graduate Text in Mathematics*; Springer: Berlin/Heidelberg, Germany, 2000; Volume 198.

 © 2019 by the authors. Licensee MDPI, Basel, Switzerland. This article is an open access article distributed under the terms and conditions of the Creative Commons Attribution (CC BY) license (http://creativecommons.org/licenses/by/4.0/).

 *symmetry*

Article

Durrmeyer-Type Generalization of Parametric Bernstein Operators

Arun Kajla [1], Mohammad Mursaleen [2,3,4,*] and Tuncer Acar [5]

[1] School of Basic Sciences, Faculty of Mathematics, Central University of Haryana, Haryana-123029, India; arunkajla@cuh.ac.in
[2] Department of Medical Research, China Medical University Hospital, China Medical University, Taichung 40402, Taiwan
[3] Department of Mathematics, Aligarh Muslim University, Aligarh 202002, India
[4] Department of Computer Science and Information Engineering, Asia University, Taichung 41354, Taiwan
[5] Department of Mathematics, Faculty of Science, Selcuk University, Selcuklu, Konya 42003, Turkey; tuncer.acar@selcuk.edu.tr
* Correspondence: mursaleen.mm@amu.ac.in; Tel.: +91-941-149-1600

Received: 13 June 2020; Accepted: 6 July 2020; Published: 8 July 2020

Abstract: In this paper, we present a Durrmeyer type generalization of parametric Bernstein operators. Firstly, we study the approximation behaviour of these operators including a local and global approximation results and the rate of approximation for the Lipschitz type space. The Voronovskaja type asymptotic formula and the rate of convergence of functions with derivatives of bounded variation are established. Finally, the theoretical results are demonstrated by using MAPLE software.

Keywords: Bernstein operators; rate of approximation; Voronovskaja type asymptotic formula

1. Introduction

A first fundamental result in approximation theory was Weierstrass approximation theorem [1] which forms the solid foundation of Approximation Theory. The proof of the theorem was quite long and difficult. So there were several proofs given by different famous mathematicians. One of them was given by Bernstein [2] which was easy and elegant, which also motivated the researchers to construct operators to deal with the approximation problems in different settings. Here, we give a Durrmeyer type generalization of parametric Bernstein operators. Let $C(J)$ be the space of all real valued continuous functions S on the interval $J = [0,1]$. For $S \in C(J)$, Chen et al. [3] introduced a new family of generalized Bernstein operators depending upon a non-negative real parameter $0 \leq \theta \leq 1$, which is given as follows:

$$T_m^{(\theta)}(S;x) = \sum_{s=0}^{m} p_{m,s}^{(\theta)}(x) S\left(\frac{s}{m}\right), \quad x \in J, \tag{1}$$

where

$$p_{m,s}^{(\theta)}(x) = \left[\binom{m-2}{s}(1-\theta)x + \binom{m-2}{s-2}(1-\theta)(1-x) + \binom{m}{s}\theta x(1-x)\right] x^{s-1}(1-x)^{m-s-1},$$

$m \geq 2$, $p_{1,0}^{(\theta)}(x) = 1 - x$, $p_{1,1}^{(\theta)}(x) = x$. For $\theta = 1$, it reduces to original Bernstein operators.

Several types of such operators have been studied so far, for example, Kajla and Acar [4] gave the integral variant of the operators (1) and studied the approximation properties of these operators. Genuine Bernstein–Durrmeyer type operators were defined and studied in [5]. Abel and Heilmann [6] studied the complete asymptotic expansion of the Bernstein–Durrmeyer operators. Cárdenas–Morales and Gupta [7] considered a two-parameter family of summation-integral type operators involving Pólya–Eggenberger distribution. In 2015, Abel et al. [8] presented the Durrmeyer type modification of the Stancu operators and obtained some approximation theorems. Agrawal et al. [9] defined Stancu type Kantorovich modification of q-Bernstein-Schurer operators and studied some approximation theorems on uniform convergence as well as A-statistical convergence. Ansari et al. in [10] proposed Jakimovski–Leviatan–Durrmeyer type operators based on Appell polynomials and obtained some approximation results, e.g.,Voronovskaja type asymptotic formula, rate of convergence and weighted approximation of these operators. Acar et al. [11] presented a general class of linear positive operators and established Voronovskaya type theorems. In 2019, Mursaleen et al. [12] considered Stancu–Jakimovski–Leviatan–Durrmeyer type operators and studied simultaneous approximation and A-statistical approximation properties of these operators.

Acu and Kajla [13] established θ-Bernstein operators depend on parameters $\rho_1, \rho_2 \in \mathbb{N} \cup \{0\}$ as follows:

$$\mathcal{B}_{m,\rho_1,\rho_2}^{(\theta)}(\mathcal{S};x) = \sum_{\mu=0}^{m-\rho_1\rho_2} p_{m-\rho_1\rho_2,\mu}^{(\theta)}(x) \sum_{s=0}^{\rho_2} p_{\rho_2,s}^{(\theta)}(x)\mathcal{S}\left(\frac{\mu+s\rho_1}{m}\right). \tag{2}$$

If $\rho_1 = \rho_2 = 0$, these operators reduces to the operators $T_m^{(\theta)}$.

For $\mathcal{S} \in C(\jmath)$, we introduce a Durrmeyer type modification of the operators (2) as follows:

$$U_{m,\rho_1,\rho_2}^{(\theta)}(\mathcal{S};x) = (m+1)\sum_{\mu=0}^{m-\rho_1\rho_2} p_{m-\rho_1\rho_2,\mu}^{(\theta)}(x) \sum_{s=0}^{\rho_2} p_{\rho_2,s}^{(\theta)}(x) \int_0^1 p_{m,\mu+s\rho_1}(t)\mathcal{S}(t)dt. \tag{3}$$

The aim of this paper is to derive approximation properties for the operators (3) by working on Korovkin's results [14]. We also compute the rate of convergence involving modulus of smoothness and Lipschitz type function.

2. Auxiliary Results

In this section, we derive some auxiliary results which will be used in proving our main results of subsequent sections. First, we determine moments and central moments for the operators (3).

Lemma 1. *Let* $e_i(t) = t^i, i = 0, 1, 2 \cdots$. *For the operators* $U_{m,\rho_1,\rho_2}^{(\theta)}$, *we have*

(i) $U_{m,\rho_1,\rho_2}^{(\theta)}(e_0;x) = 1;$

(ii) $U_{m,\rho_1,\rho_2}^{(\theta)}(e_1;x) = \dfrac{mx+1}{(m+2)};$

(iii) $U_{m,\rho_1,\rho_2}^{(\theta)}(e_2;x) = \dfrac{x^2\left(m^2 - m - 2 + \rho_1\rho_2 - \rho_1^2(\rho_2 - 2\theta + 2) + 2\theta\right)}{(m+3)(m+2)}$

$\qquad + \dfrac{x\left(2 + 4m - \rho_1\rho_2 + \rho_1^2(\rho_2 - 2\theta + 2) - 2\theta\right)}{(m+3)(m+2)} + \dfrac{2}{(m+3)(m+2)};$

(iv) $U_{m,\rho_1,\rho_2}^{(\theta)}(e_3;x) = \dfrac{x^3}{(m+4)(m+3)(m+2)}\left[m^3 - 3m^2 - 4m + \rho_1\rho_2(3m-2) + 2\rho_1^3(\rho_2 - 6\theta + 6)\right.$

$\qquad\left. - 3m\rho_1^2(2 + \rho_2 - 2\theta) - 12(\theta - 1) + 6m\theta\right]$

$\qquad + \dfrac{x^2}{(m+4)(m+3)(m+2)}\left[9m^2 + 9\rho_1\rho_2 - 3m - 3m\rho_1\rho_2 - 3\rho_1^3(6 + \rho_2 - 6\theta) - 6\rho_1^2(2 + \rho_2 - 2\theta)\right.$

$\qquad\left. + 3m\rho_1^2(2 + \rho_2 - 2\theta) + 30(\theta - 1) - 6m\theta\right]$

$\qquad + \dfrac{x\left(18m - 7\rho_1\rho_2 + \rho_1^3(\rho_2 - 6\theta + 6) + 6\rho_1^2(2 + \rho_2 - 2\theta) - 18(\theta - 1)\right)}{(m+4)(m+3)(m+2)} + \dfrac{6}{(m+4)(m+3)(m+2)};$

(v) $U_{m,\rho_1,\rho_2}^{(\theta)}(e_4;x) = \dfrac{x^4}{(m+5)(m+4)(m+3)(m+2)}\left[54m - m^2 - 6m^3 + m^4 - 6\rho_1\rho_2 - 14m\rho_1\rho_2 + 3\rho_1^2\rho_2^2 - 6\rho_1^3\rho_2^2 + \right.$

$\qquad 3\rho_1^4\rho_2^2 - 6m^2\rho_1(-\rho_2 + \rho_1(2 + \rho_2)) + 8m\rho_1^3(6 + \rho_2 - 6\theta) + 6\rho_1^4\rho_2(1 - 2\theta) + 6m\rho_1^2(2 + \rho_2 - 2\theta) + 72(-1 + \theta) + $

$$72\rho_1^4(-1+\theta) - 12\rho_1^2\rho_2(-1+\theta) + 12\rho_1^3\rho_2(-1+\theta) + 24\rho_1^2(-1+\theta)^2 - 60m\theta + 12m^2\left(1+\rho_1^2\right)\theta + 12\rho_1\rho_2\theta\Big]$$

$$+ \frac{x^3}{(m+5)(m+4)(m+3)(m+2)}\Big[16m^3 - 124m - 36m^2 - 8\rho_1\rho_2 + 54m\rho_1\rho_2 + 32\rho_1^3\rho_2 - 6\rho_1^2\rho_2^2 + 12\rho_1^3\rho_2^2 -$$
$$6\rho_1^4\rho_2^2 + 6m^2\rho_1(-\rho_2 + \rho_1(2+\rho_2)) - 12m\rho_1^3(6+\rho_2 - 6\theta) - 12\rho_1^4\rho_2(1-2\theta) - 42m\rho_1^2(2+\rho_2 - 2\theta) - 264(-1+$$
$$\theta) - 120\rho_1^3(-1+\theta) - 144\rho_1^4(-1+\theta) + 24\rho_1^2\rho_2(-1+\theta) - 12\rho_1^3\rho_2(-1+\theta) - 48\rho_1^2(-1+\theta)^2 + 156m\theta -$$
$$12m^2\left(1+\rho_1^2\right)\theta - 24\rho_1\rho_2\theta - 12\rho_1^3\rho_2\theta\Big] + \frac{x^2}{(m+5)(m+4)(m+3)(m+2)}\Big[24m + 72m^2 + 60\rho_1\rho_2 - 40m\rho_1\rho_2 -$$
$$35\rho_1^2\rho_2 - 42\rho_1^3\rho_2 - \rho_1^4\rho_2 + 3\rho_1^2\rho_2^2 - 6\rho_1^3\rho_2^2 + 3\rho_1^4\rho_2^2 + 4m\rho_1^3(6+\rho_2 - 6\theta) + 6\rho_1^4\rho_2(1-2\theta) + 36m\rho_1^2(2+\rho_2 - 2\theta) +$$
$$336(-1+\theta) + 70\rho_1^2(-1+\theta) + 180\rho_1^3(-1+\theta) + 86\rho_1^4(-1+\theta) - 12\rho_1^2\rho_2(-1+\theta) + 24\rho_1^2(-1+\theta)^2 - 96m\theta +$$
$$12\rho_1\rho_2\theta + 12\rho_1^3\rho_2\theta\Big]$$

$$+ \frac{x\left(96m - 46\rho_1\rho_2 + 35\rho_1^2\rho_2 + 10\rho_1^3\rho_2 + \rho_1^4\rho_2 - 144(-1+\theta) - 70\rho_1^2(-1+\theta) - 60\rho_1^3(-1+\theta) - 14\rho_1^4(-1+\theta)\right)}{(m+5)(m+4)(m+3)(m+2)}$$

$$+ \frac{24}{(2+m)(3+m)(4+m)(5+m)}.$$

Let $\Theta_{m,\rho_1,\rho_2}^{(\theta),m} := U_{m,\rho_1,\rho_2}^{(\theta)}((t-x)^m; x)$, $m = 1, 2, 4$ be the central moments of $U_{m,\rho_1,\rho_2}^{(\theta)}$.

Lemma 2. *For the operators* $U_{m,\rho_1,\rho_2}^{(\theta)}$, *we get*

(i) $\Theta_{m,\rho_1,\rho_2}^{(\theta),1}(x) = \left(\dfrac{1-2x}{2+m}\right)$;

(ii) $\Theta_{m,\rho_1,\rho_2}^{(\theta),2}(x) = \dfrac{2}{(2+m)(3+m)} + \dfrac{x(2m - \rho_1(\rho_2 - \rho_1(2+\rho_2 - 2\theta)) - 2(2+\theta))}{(2+m)(3+m)}$
$+ \dfrac{x^2(-2m + \rho_1(\rho_2 - \rho_1(2+\rho_2 - 2\theta)) + 2(2+\theta))}{(2+m)(3+m)}.$

Lemma 3. *For* $m \in \mathbb{N}$, *we have*

$$U_{m,\rho_1,\rho_2}^{(\theta)}((t-x)^2; x) \leq \frac{\mathcal{W}_{\rho_1,\rho_2}^{(\theta)}x(1-x)}{(m+2)} = \delta_{m,\rho_1,\rho_2}^{(\theta)}(x), \forall x \in \jmath,$$

where $\mathcal{W}_{\rho_1,\rho_2}^{(\theta)}$ *is a positive constant depending on* ρ_1, ρ_2 *and* θ.

Proof. This lemma is established by direct computation and the details are missing. $\square$

Remark 1. *For the operators* $U_{m,\rho_1,\rho_2}^{(\theta)}$, *we get*

$$\lim_{m\to\infty} m\,\Theta_{m,\rho_1,\rho_2}^{(\theta),1}(x) = (1-2x),$$
$$\lim_{m\to\infty} m\,\Theta_{m,\rho_1,\rho_2}^{(\theta),2}(x) = 2x(1-x),$$
$$\lim_{m\to\infty} m^2\,\Theta_{m,\rho_1,\rho_2}^{(\theta),4}(x) = 12x^2(1-x)^2.$$

Lemma 4. *For* $\mathcal{S} \in C(\jmath)$, *we have*
$$\|U_{m,\rho_1,\rho_2}^{(\theta)}(\mathcal{S}; x)\| \leq \|\mathcal{S}\|.$$

Proof. From Lemma 1 and Equation (3), we obtain

$$\|U_{m,\rho_1,\rho_2}^{(\theta)}\| \leq (m+1)\sum_{\mu=0}^{m-\rho_1\rho_2} p_{m-\rho_1\rho_2,\mu}^{(\theta)}(x)\sum_{s=0}^{\rho_2} p_{\rho_2,s}^{(\theta)}(x)\int_0^1 p_{m,\mu+s\rho_1}(t)|\mathcal{S}(t)|dt$$

$$\leq \|\mathcal{S}\|U_{m,\rho_1,\rho_2}^{(\theta)}(e_0; x) = \|\mathcal{S}\|.$$

$\square$

Theorem 1. *Suppose that $\mathcal{S} \in C(\jmath)$. Show that $\lim\limits_{m\to\infty} U_{m,\rho_1,\rho_2}^{(\theta)}(\mathcal{S};x) = \mathcal{S}(x)$, uniformly in $\jmath$.*

Proof. Since $U_{m,\rho_1,\rho_2}^{(\theta)}(1;x) = 1$, $U_{m,\rho_1,\rho_2}^{(\theta)}(t;x) \to x$, $U_{m,\rho_1,\rho_2}^{(\theta)}(t^2;x) \to x^2$ as $m \to \infty$, uniformly in $\jmath$. By Korovkin's results, it follows that $U_{m,\rho_1,\rho_2}^{(\theta)}(\mathcal{S};x)$ converges to $\mathcal{S}(x)$ uniformly on $\jmath$. $\square$

3. Voronovskaja Type Theorems

Here, we establish the Voronovskaja, Grüss-Voronovskaja type theorems and related results.

Theorem 2. *Suppose that $\mathcal{S} \in C(\jmath)$. If $\mathcal{S}', \mathcal{S}''$ exists at a point $x \in \jmath$ then*

$$\lim_{m\to\infty} m\left(U_{m,\rho_1,\rho_2}^{(\theta)}(\mathcal{S};x) - \mathcal{S}(x) \right) = (1 - 2x)\mathcal{S}'(x) + x(1 - x)\mathcal{S}''(x), \tag{4}$$

Further, if $\mathcal{S}'' \in C(\jmath)$ then (4) holds uniformly in $\jmath$.

Proof. Applying the application of Taylor's theorem, we have

$$\mathcal{S}(t) = \mathcal{S}(x) + (t - x)\mathcal{S}'(x) + \frac{1}{2}(t - x)^2 \mathcal{S}''(x) + \rho_1(t,x)(t - x)^2, \tag{5}$$

where $\rho_1(t,x) \to 0$ as $t \to x$ and is a continuous function on $\jmath$. Applying $U_{m,\rho_1,\rho_2}^{(\theta)}$ to (5), we get

$$U_{m,\rho_1,\rho_2}^{(\theta)}(\mathcal{S};x) - \mathcal{S}(x) = \mathcal{S}'(x)U_{m,\rho_1,\rho_2}^{(\theta)}((t-x);x) + \frac{1}{2}\mathcal{S}''(x)U_{m,\rho_1,\rho_2}^{(\theta)}((t-x)^2;x) + U_{m,\rho_1,\rho_2}^{(\theta)}(\rho_1(t,x)(t-x)^2;x),$$

$$\lim_{m\to\infty} m\left(U_{m,\rho_1,\rho_2}^{(\theta)}(\mathcal{S};x) - \mathcal{S}(x) \right) = (1 - 2x)\mathcal{S}'(x) + x(1 - x)\mathcal{S}''(x) + \lim_{m\to\infty} mU_{m,\rho_1,\rho_2}^{(\theta)}(\rho_1(t,x)(t-x)^2;x).$$

Since $\rho_1(t,x) \to 0$ as $t \to x$, for a given $\epsilon > 0$, there exists $\delta > 0$ such that $|\rho_1(t,x)| < \epsilon$ whenever $|t - x| < \delta$. For $|t - x| \geq \delta$, we have $|\rho_1(t,x)| \leq M\frac{(t-x)^2}{\delta^2}$, for some $M > 0$. Let $\chi_\delta(t)$ denote the characteristic function of the interval $(x - \delta, x + \delta)$. In view of Remark 1, we have

$$
\begin{aligned}
|U_{m,\rho_1,\rho_2}^{(\theta)}(\rho_1(t,x)(t-x)^2;x)| &\leq U_{m,\rho_1,\rho_2}^{(\theta)}(|\rho_1(t,x)|(t-x)^2\chi_\delta(t);x) + U_{m,\rho_1,\rho_2}^{(\theta)}(|\rho_1(t,x)|(t-x)^2(1-\chi_\delta(t));x) \\
&< \epsilon\, U_{m,\rho_1,\rho_2}^{(\theta)}((t-x)^2;x) + \frac{M}{\delta^2}U_{m,\rho_1,\rho_2}^{(\theta)}((t-x)^4;x) \\
&= \epsilon\, O\left(\frac{1}{m}\right) + O\left(\frac{1}{m^2}\right).
\end{aligned}
$$

which implies that $\lim\limits_{m\to\infty} mU_{m,\rho_1,\rho_2}^{(\theta)}(\rho_1(t,x)(t-x)^2;x) = 0$, due to the arbitrariness of $\epsilon > 0$. This complete the first half of the theorem.

To show the uniformity postulation, by the definition of uniformly continuity of $\mathcal{S}$ in $\jmath$, the δ must be independent of x and all the other estimates hold uniformly in $x \in \jmath$. $\square$

In [15], Acar et al. obtained a Grüss type approximation result and a Grüss-Voronovskaja-type result for linear and positive operators. Many authors have established in this direction so that we refer the authors to [16–18] and references therein.

The next result is the Grüss–Voronovskaja type theorem for $U_{m,\rho_1,\rho_2}^{(\theta)}$.

Theorem 3. *Let $\mathcal{S}, \hbar \in C^2(\jmath)$. Then, for each $x \in \jmath$,*

$$\lim_{m\to\infty} m\left\{ U_{m,\rho_1,\rho_2}^{(\theta)}((\mathcal{S}\hbar);x) - U_{m,\rho_1,\rho_2}^{(\theta)}(\mathcal{S};x)U_{m,\rho_1,\rho_2}^{(\theta)}(\hbar;x) \right\} = \mathcal{S}'(x)\hbar'(x)2x(1 - x).$$

Proof. The following relation holds

$$
\begin{aligned}
&U^{(\theta)}_{m,\rho_1,\rho_2}(\mathcal{S}\hbar;x) - U^{(\theta)}_{m,\rho_1,\rho_2}(\mathcal{S};x)U^{(\theta)}_{m,\rho_1,\rho_2}(\hbar;x) = U^{(\theta)}_{m,\rho_1,\rho_2}(\mathcal{S}\hbar;x) - \mathcal{S}(x)\hbar(x) - (\mathcal{S}\hbar)'(x)\Theta^{(\theta),1}_{m,\rho_1,\rho_2}(x) - \frac{1}{2}(\mathcal{S}\hbar)''(x)\Theta^{(\theta),2}_{m,\rho_1,\rho_2}(x) \\
&- \hbar(x)\left\{ U^{(\theta)}_{m,\rho_1,\rho_2}(\mathcal{S};x) - \mathcal{S}(x) - \mathcal{S}'(x)\Theta^{(\theta),1}_{m,\rho_1,\rho_2}(x) - \frac{1}{2}\mathcal{S}''(x)\Theta^{(\theta),2}_{m,\rho_1,\rho_2}(x) \right\} \\
&- U^{(\theta)}_{m,\rho_1,\rho_2}(\mathcal{S};x)\left\{ U^{(\theta)}_{m,\rho_1,\rho_2}(\hbar;x) - \hbar(x) - \hbar'(x)\Theta^{(\theta),1}_{m,\rho_1,\rho_2}(x) - \frac{1}{2}\hbar''(x)\Theta^{(\theta),2}_{m,\rho_1,\rho_2}(x) \right\} \\
&+ \frac{1}{2}\Theta^{(\theta),2}_{m,\rho_1,\rho_2}(x)\left\{ \mathcal{S}(x)\hbar''(x) + 2\mathcal{S}'(x)\hbar'(x) - \hbar''(x)U^{(\theta)}_{m,\rho_1,\rho_2}(\mathcal{S};x) \right\} + \Theta^{(\theta),1}_{m,\rho_1,\rho_2}(x)\left\{ \mathcal{S}(x)\hbar'(x) - \hbar'(x)U^{(\theta)}_{m,\rho_1,\rho_2}(\mathcal{S};x) \right\}.
\end{aligned}
$$

Now, by using Theorem 1, Theorem 2 and Remark 1, we get

$$
\begin{aligned}
&\lim_{m\to\infty} m\left\{ U^{(\theta)}_{m,\rho_1,\rho_2}(\mathcal{S}\hbar;x) - U^{(\theta)}_{m,\rho_1,\rho_2}(\mathcal{S};x)U^{(\theta)}_{m,\rho_1,\rho_2}(\hbar;x) \right\} \\
&= \lim_{m\to\infty} m\mathcal{S}'(x)\hbar'(x)\Theta^{(\theta),2}_{m,\rho_1,\rho_2}(x) + \lim_{m\to\infty} \frac{1}{2}m\hbar''(x)\left\{ \mathcal{S}(x) - U^{(\theta)}_{m,\rho_1,\rho_2}(\mathcal{S};x) \right\}\Theta^{(\theta),2}_{m,\rho_1,\rho_2}(x) \\
&+ \lim_{m\to\infty} m\hbar'(x)\left\{ \mathcal{S}(x) - U^{(\theta)}_{m,\rho_1,\rho_2}(\mathcal{S};x) \right\}\Theta^{(\theta),1}_{m,\rho_1,\rho_2}(x) = \mathcal{S}'(x)\hbar'(x)2x(1-x).
\end{aligned}
$$

$\square$

Lipschitz-type space with two parameters $\alpha_1 \geq 0, \alpha_2 > 0$ is defined in [19] as below:

$$
Lip^{(\alpha_1,\alpha_2)}_M(\sigma) := \left\{ \mathcal{S} \in C(\jmath) : |\mathcal{S}(t) - \mathcal{S}(x)| \leq M\frac{|t-x|^\sigma}{(t+\alpha_1 x^2 + \alpha_2 x)^{\frac{\sigma}{2}}}; t \in \jmath, x \in (0,1] \right\},
$$

where $0 < \sigma \leq 1$.

Theorem 4. *Suppose that* $\mathcal{S} \in Lip^{(\alpha_1,\alpha_2)}_M(\sigma)$. *Prove that*

$$
\left| U^{(\theta)}_{m,\rho_1,\rho_2}(\mathcal{S};x) - \mathcal{S}(x) \right| \leq M\left(\frac{\Theta^{(\theta),2}_{m,\rho_1,\rho_2}(x)}{\alpha_1 x^2 + \alpha_2 x} \right)^{\sigma/2}, \forall x \in (0,1].
$$

Proof. Using the application of Holder's inequality and Lemma 2, we may write

$$
\begin{aligned}
\left| U^{(\theta)}_{m,\rho_1,\rho_2}(\mathcal{S};x) - \mathcal{S}(x) \right| &\leq (m+1)\sum_{\mu=0}^{m-\rho_1\rho_2} p^{(\theta)}_{m-\rho_1\rho_2,\mu}(x)\sum_{s=0}^{\rho_2} p^{(\theta)}_{\rho_2,s}(x)\int_0^1 |\mathcal{S}(t) - \mathcal{S}(x)|\, p_{m,\mu+s\rho_1}(t)dt \\
&\leq (m+1)\sum_{\mu=0}^{m-\rho_1\rho_2} p^{(\theta)}_{m-\rho_1\rho_2,\mu}(x)\sum_{s=0}^{\rho_2} p^{(\theta)}_{\rho_2,s}(x)\left(\int_0^1 |\mathcal{S}(t) - \mathcal{S}(x)|^{\frac{2}{\sigma}} p_{m,\mu+s\rho_1}(t)dt \right)^{\frac{\varsigma}{2}} \\
&\leq \left\{ (m+1)\sum_{\mu=0}^{m-\rho_1\rho_2} p^{(\theta)}_{m-\rho_1\rho_2,\mu}(x)\sum_{s=0}^{\rho_2} p^{(\theta)}_{\rho_2,s}(x)\int_0^1 |\mathcal{S}(t) - \mathcal{S}(x)|^{\frac{2}{\sigma}} p_{m,\mu+s\rho_1}(t)dt \right\}^{\frac{\varsigma}{2}} \\
&\quad \times \left((m+1)\sum_{\mu=0}^{m-\rho_1\rho_2} p^{(\theta)}_{m-\rho_1\rho_2,\mu}(x)\sum_{s=0}^{\rho_2} p^{(\theta)}_{\rho_2,s}(x)\int_0^1 p_{m,\mu+s\rho_1}(t)dt \right)^{\frac{2-\varsigma}{2}} \\
&= \left((m+1)\sum_{\mu=0}^{m-\rho_1\rho_2} p^{(\theta)}_{m-\rho_1\rho_2,\mu}(x)\sum_{s=0}^{\rho_2} p^{(\theta)}_{\rho_2,s}(x)\int_0^1 |\mathcal{S}(t) - \mathcal{S}(x)|^{\frac{2}{\sigma}} p_{m,\mu+s\rho_1}(t)dt \right)^{\frac{\varsigma}{2}} \\
&\leq M\left((m+1)\sum_{\mu=0}^{m-\rho_1\rho_2} p^{(\theta)}_{m-\rho_1\rho_2,\mu}(x)\sum_{s=0}^{\rho_2} p^{(\theta)}_{\rho_2,s}(x)\int_0^1 \frac{(t-x)^2}{(t+\alpha_1 x^2 + \alpha_2 x)} p_{m,\mu+s\rho_1}(t)dt \right)^{\frac{\varsigma}{2}} \\
&\leq \frac{M}{(\alpha_1 x^2 + \alpha_2 x)^{\frac{\varsigma}{2}}}\left((m+1)\sum_{\mu=0}^{m-\rho_1\rho_2} p^{(\theta)}_{m-\rho_1\rho_2,\mu}(x)\sum_{s=0}^{\rho_2} p^{(\theta)}_{\rho_2,s}(x)\int_0^1 (t-x)^2 p_{m,\mu+s\rho_1}(t)dt \right)^{\frac{\varsigma}{2}} \\
&= \frac{M}{(\alpha_1 x^2 + \alpha_2 x)^{\frac{\varsigma}{2}}} U^{(\theta)}_{m,\rho_1,\rho_2}((t-x)^2;x)^{\frac{\varsigma}{2}} \\
&= \frac{M}{(\alpha_1 x^2 + \alpha_2 x)^{\frac{\varsigma}{2}}}(\Theta^{(\theta),2}_{m,\rho_1,\rho_2}(x))^{\frac{\varsigma}{2}}.
\end{aligned}
$$

□

Theorem 5. *For $\mathcal{S} \in C^1(\jmath)$ and $x \in \jmath$, we have*

$$\left| U^{(\theta)}_{m,\rho_1,\rho_2}(\mathcal{S};x) - \mathcal{S}(x) \right| \leq \left| \frac{1-2x}{(m+2)} \right| |\mathcal{S}'(x)| + 2\sqrt{\Theta^{(\theta),2}_{m,\rho_1,\rho_2}(x)}\, \omega\left(\mathcal{S}', \sqrt{\Theta^{(\theta),2}_{m,\rho_1,\rho_2}(x)} \right). \tag{6}$$

Proof. Let $\mathcal{S} \in C^1(\jmath)$. For any $t, x \in \jmath$, we have

$$\mathcal{S}(t) - \mathcal{S}(x) = \mathcal{S}'(x)(t-x) + \int_x^t \left(\mathcal{S}'(u) - \mathcal{S}'(x) \right) du.$$

Using $U^{(\theta)}_{m,\rho_1,\rho_2}(\cdot\,;x)$ on both sides of the above relation, we have

$$U^{(\theta)}_{m,\rho_1,\rho_2}(\mathcal{S}(t) - \mathcal{S}(x); q_m, x) = \mathcal{S}'(x) U^{(\theta)}_{m,\rho_1,\rho_2}(t-x;x) + U^{(\theta)}_{m,\rho_1,\rho_2}\left(\int_x^t \left(\mathcal{S}'(u) - \mathcal{S}'(x) \right) du; x \right)$$

Applying $|\mathcal{S}(t) - \mathcal{S}(x)| \leq \omega(\mathcal{S}, \delta) \left(\frac{|t-x|}{\delta} + 1 \right), \delta > 0$, we have

$$\left| \int_x^t \left(\mathcal{S}'(u) - \mathcal{S}'(x) \right) du \right| \leq \omega(\mathcal{S}', \delta) \left(\frac{(t-x)^2}{\delta} + |t-x| \right),$$

it follows that

$$\left| U^{(\theta)}_{m,\rho_1,\rho_2}(\mathcal{S};x) - \mathcal{S}(x) \right| \leq |\mathcal{S}'(x)| \ \left| U^{(\theta)}_{m,\rho_1,\rho_2}(t-x;x) \right| + \omega(\mathcal{S}', \delta) \left\{ \frac{1}{\delta} U^{(\theta)}_{m,\rho_1,\rho_2}((t-x)^2;x) + U^{(\theta)}_{m,\rho_1,\rho_2}(|t-x|;x) \right\}.$$

Applying Cauchy–Schwarz inequality, we get

$$\begin{aligned}
\left| U^{(\theta)}_{m,\rho_1,\rho_2}(\mathcal{S};x) - \mathcal{S}(x) \right| &\leq \ |\mathcal{S}'(x)| \ \left| U^{(\theta)}_{m,\rho_1,\rho_2}(t-x;x) \right| \\
&\quad + \omega(\mathcal{S}', \delta) \left\{ \frac{1}{\delta} \sqrt{U^{(\theta)}_{m,\rho_1,\rho_2}((t-x)^2;x)} + 1 \right\} \sqrt{U^{(\theta)}_{m,\rho_1,\rho_2}((t-x)^2;x)}.
\end{aligned}$$

Now, taking $\delta = \sqrt{\Theta^{(\theta),2}_{m,\rho_1,\rho_2}(x)}$, we get (6). □

4. Local Approximation

In this section, we study the local approximation property for our operators with the help of K-functional.

The K-functional is given by:

$$K_2(\mathcal{S}, \delta) = \inf\{ \|\mathcal{S} - \hbar\| + \delta \|\hbar''\| : \hbar \in W^2 \} \ (\delta > 0),$$

where $W^2 = \{\hbar : \hbar'' \in C(\jmath)\}$ and uniform norm on $C(\jmath)$ is denoted by $\|.\|$. By [20] there will be a positive constant $M > 0$ such that

$$K_2(\mathcal{S}, \delta) \leq M\omega_2(\mathcal{S}, \sqrt{\delta}), \tag{7}$$

where the second order modulus of continuity for $\mathcal{S} \in C(\jmath)$ is defined as

$$\omega_2(\mathcal{S}, \sqrt{\delta}) = \sup_{0 < h \leq \sqrt{\delta}} \sup_{x, x+2h \in \jmath} |\mathcal{S}(x+2h) - 2\mathcal{S}(x+h) + \mathcal{S}(x)|.$$

We define the usual modulus of continuity for $S \in C(\jmath)$ as

$$\omega(S, \delta) = \sup_{0 < h \leq \delta} \sup_{x, x+h \in \jmath} |S(x+h) - S(x)|.$$

Theorem 6. *For the operators $U^{(\theta)}_{m,\rho_1,\rho_2}$, there exists a constant $M > 0$ such that*

$$\left| U^{(\theta)}_{m,\rho_1,\rho_2}(S; x) - S(x) \right| \leq M\omega_2\left(S, (m+2)^{-1/2}\sqrt{\delta^{(\theta)}_{m,\rho_1,\rho_2}(x)} \right) + \omega\left(S, \left| \frac{1-2x}{m+2} \right| \right),$$

where $S \in C(\jmath)$, $\theta \in \jmath$, $\delta^{(\theta)}_{m,\rho_1,\rho_2}(x) = \varphi^2(x) + \frac{1}{(m+2)}$ and $x \in \jmath$.

Proof. We define the auxiliary operators as follows:

$$\overline{U}^{(\theta)}_{m,\rho_1,\rho_2}(S; x) = U^{(\theta)}_{m,\rho_1,\rho_2}(S; x) + S(x) - S\left(\frac{mx+1}{m+2} \right).$$

Then, we can easily check that

$$\overline{U}^{(\theta)}_{m,\rho_1,\rho_2}(1; x) = 1 \quad and \quad \overline{U}^{(\theta)}_{m,\rho_1,\rho_2}(t; x) = x.$$

By the application of Taylor's theorem and taking $t \in \jmath$ and $\hbar \in W^2$, we get

$$\hbar(t) = \hbar(x) + (t-x)\hbar'(x) + \int_x^t (t-u)\hbar''(u)du.$$

The operator $\overline{U}^{(\theta)}_{m,\rho_1,\rho_2}$ is applied in the above equation on both sides, we obtain

$$\overline{U}^{(\theta)}_{m,\rho_1,\rho_2}(\hbar; x) = \hbar(x) + \overline{U}^{(\theta)}_{m,\rho_1,\rho_2}\left(\int_x^t (t-u)\hbar''(u)du \right)$$

$$= \hbar(x) + U^{(\theta)}_{m,\rho_1,\rho_2}\left(\int_x^t (t-u)\hbar''(u)du, x \right) - \int_x^{\frac{mx+1}{(m+2)}} \left(\frac{mx+1}{m+2} - u \right) \hbar''(u)du.$$

Hence

$$\left| \overline{U}^{(\theta)}_{m,\rho_1,\rho_2}(\hbar; x) - \hbar(x) \right| \leq U^{(\theta)}_{m,\rho_1,\rho_2}\left(\left| \int_x^t |t-u||\hbar''(u)|du \right|, x \right) + \left| \int_x^{\frac{mx+1}{(m+2)}} \left| \frac{mx+1}{m+2} - u \right| |\hbar''(u)|du \right|$$

$$\leq \left\{ U^{(\theta)}_{m,\rho_1,\rho_2}((t-x)^2; x) + \left(\frac{mx+1}{m+2} - x \right)^2 \right\} \|\hbar''\|$$

$$= \left\{ U^{(\theta)}_{m,\rho_1,\rho_2}((t-x)^2; x) + \left(\frac{1-2x}{m+2} \right)^2 \right\} \|\hbar''\|. \tag{8}$$

From Lemma 3, we have

$$U^{(\theta)}_{m,\rho_1,\rho_2}((t-x)^2; x) + \left(\frac{1-2x}{m+2} \right)^2 \leq \frac{2}{(m+2)}\delta^{(\theta)}_{m,\rho_1,\rho_2}(x) + \left(\frac{1-2x}{m+2} \right)^2$$

$$\leq \frac{2}{(m+2)}\delta^{(\theta)}_{m,\rho_1,\rho_2}(x) + \frac{1}{(m+2)^2}$$

$$\leq \frac{3}{(m+2)}\delta^{(\theta)}_{m,\rho_1,\rho_2}(x). \tag{9}$$

Thus, by (8) we have

$$| \overline{U}_{m,\rho_1,\rho_2}^{(\theta)}(\hbar;x) - \hbar(x) | \leq \frac{3}{(m+2)} \delta_{m,\rho_1,\rho_2}^{(\theta)}(x) ||\hbar''||, \tag{10}$$

where $x \in \jmath$. Furthermore, by Lemma 4, we have

$$| \overline{U}_{m,\rho_1,\rho_2}^{(\theta)}(\mathcal{S};x) | \leq 3||\mathcal{S}||, \tag{11}$$

for all $\mathcal{S} \in C(\jmath)$ and $x \in \jmath$.

Now, for $\mathcal{S} \in C(\jmath)$ and $\hbar \in W^2$, using (10) and (11) we obtain that

$$
\begin{aligned}
| U_{m,\rho_1,\rho_2}^{(\theta)}(\mathcal{S};x) - \mathcal{S}(x) | &\leq \left| \overline{U}_{m,\rho_1,\rho_2}^{(\theta)}(\mathcal{S};x) - \mathcal{S}(x) + \mathcal{S}\left(\frac{mx+1}{m+2}\right) - \mathcal{S}(x) \right| \\
&\leq |\overline{U}_{m,\rho_1,\rho_2}^{(\theta)}(\mathcal{S}-\hbar;x)| + |\overline{U}_{m,\rho_1,\rho_2}^{(\theta)}(\hbar;x) - \hbar(x)| + |\hbar(x) - \mathcal{S}(x)| \\
&\quad + \left| \mathcal{S}\left(\frac{mx+1}{m+2}\right) - \mathcal{S}(x) \right| \\
&\leq 4||\mathcal{S}-\hbar|| + \frac{3}{(m+2)} \delta_{m,\rho_1,\rho_2}^{(\theta)}(x) ||\hbar''|| + \omega\left(\mathcal{S}, \left|\frac{1-2x}{m+2}\right|\right).
\end{aligned}
$$

Using the property of infimum on the right hand side over all $\hbar \in W^2$, we have

$$| U_{m,\rho_1,\rho_2}^{(\theta)}(\mathcal{S};x) - \mathcal{S}(x) | \leq 4K_2\left(\mathcal{S}, \frac{1}{(m+2)} \delta_{m,\rho_1,\rho_2}^{(\theta)}(x)\right) + \omega\left(\mathcal{S}, \left|\frac{1-2x}{m+2}\right|\right).$$

Now by examining the relation (7), we get

$$| U_{m,\rho_1,\rho_2}^{(\theta)}(\mathcal{S};x) - \mathcal{S}(x) | \leq M\omega_2\left(\mathcal{S}, (m+2)^{-1/2}\sqrt{\delta_{m,\rho_1,\rho_2}^{(\theta)}(x)}\right) + \omega\left(\mathcal{S}, \left|\frac{1-2x}{m+2}\right|\right).$$

$\square$

5. Global Approximation

The following result provides the global approximation using the modulus of continuity of Ditzian–Totik and the related K-functional.

Suppose that $\mathcal{S} \in C(\jmath)$ and $\varphi(x)$ is defined as $\sqrt{x(1-x)}, x \in \jmath$. The second order modulus of continuity which is given by Ditzian-Totik

$$\omega_2^{\varphi}(\mathcal{S}, \sqrt{\delta}) = \sup_{0<h\leq\sqrt{\delta}} \sup_{x\pm h\varphi(x)\in\jmath} | \mathcal{S}(x + h\varphi(x)) - 2\mathcal{S}(x) + \mathcal{S}(x - h\varphi(x)) |,$$

and related K-functional is defined as,

$$\tilde{K}_{2,\varphi(x)}(\mathcal{S}, \delta) = \inf\{||\mathcal{S}-\hbar|| + \delta||\varphi^2\hbar''|| + \delta^2||\hbar''|| : \hbar \in W^2(\varphi)\}, (\delta > 0),$$

where $W^2(\varphi) = \{\hbar \in C(\jmath) : \hbar' \in AC_{loc}\jmath, \varphi^2\hbar'' \in C(\jmath)\}$ and $\hbar' \in AC_{loc}\jmath$ means that $\hbar$ is derivable and $\hbar'$ is absolutely continuous on every closed interval $[a,b] \subset (0,1)$. By ([21],Theorem 1.3.1) we can say that $\exists M > 0$, such that

$$\tilde{K}_{2,\varphi(x)}(\mathcal{S}, \delta) \leq M\omega_2^{\varphi}(\mathcal{S}, \sqrt{\delta}). \tag{12}$$

The first order Ditzian–Totik modulus is defined as

$$\overrightarrow{\omega_\psi}(\mathcal{S},\delta) = \sup_{0<h\le\delta} \sup_{x\pm\frac{h}{2}\psi(x)\in J} \left| \mathcal{S}\left(x+\frac{h}{2}\psi(x)\right) - \mathcal{S}\left(x-\frac{h}{2}\psi(x)\right)\right|,$$

where $\psi : J \to \mathbb{R}$ is an admissible step-weight function.

Now we state our next main theorem.

Theorem 7. *Let $\mathcal{S} \in C(J)$ and $0 \le \theta \le 1$. Then, for $x \in J$,*

$$||U^{(\theta)}_{m,\rho_1,\rho_2}\mathcal{S} - \mathcal{S}|| \le M\omega_2^\varphi(\mathcal{S},(m+2)^{-1/2}) + \overrightarrow{\omega_\psi}\left(\mathcal{S},(m+2)^{-1}\right) + \omega\left(\mathcal{S};(m+2)^{-1}\right),$$

where $\varphi^2(x) = x(1-x)$ and $\psi(x) = \begin{cases} 1-2x & x \in [0,1/2] \\ 2x-1 & x \in [1/2,1] \end{cases}$.

Proof. The auxiliary operators is considered as

$$\overline{U}^{(\theta)}_{m,\rho_1,\rho_2}(\mathcal{S};x) = U^{(\theta)}_{m,\rho_1,\rho_2}(\mathcal{S};x) + \mathcal{S}(x) - \mathcal{S}\left(\frac{mx+1}{m+2}\right).$$

Let $\hbar \in W^2(\varphi)$ then by expanding $\hbar$ using Taylor's theorem and as given in the proof of Theorem 6, we get

$$|\overline{U}^{(\theta)}_{m,\rho_1,\rho_2}(\hbar;x) - \hbar(x)| \le U^{(\theta)}_{m,\rho_1,\rho_2}\left(\left|\int_x^t |t-u||\hbar''(u)|du\right|,x\right) + \int_x^{\frac{mx+1}{(m+2)}} \left|\frac{mx+1}{m+2}-u\right| |\hbar''(u)|du. \quad (13)$$

Setting $u = \beta x + (1-\beta)t, \beta \in J$, and also applying the concavity of $\delta^{(\theta)}_{m,\rho_1,\rho_2}$, we have

$$\frac{|t-u|}{\delta^{(\theta)}_{m,\rho_1,\rho_2}(u)} = \frac{\beta|t-x|}{\delta^{(\theta)}_{m,\rho_1,\rho_2}(\beta x+(1-\beta)t)} \le \frac{\beta|t-x|}{\delta^{(\theta)}_{m,\rho_1,\rho_2}(x)\beta + \delta^{(\theta)}_{m,\rho_1,\rho_2}(t)(1-\beta)} \le \frac{|t-x|}{\delta^{(\theta)}_{m,\rho_1,\rho_2}(x)}. \quad (14)$$

Thus, using (14) in the inequality (13)

$$|\overline{U}^{(\theta)}_{m,\rho_1,\rho_2}(\hbar;x) - \hbar(x)| \le U^{(\theta)}_{m,\rho_1,\rho_2}\left(\left|\int_x^t \frac{|t-u|}{\delta^{(\theta)}_{m,\rho_1,\rho_2}(u)}du\right|,x\right) ||\delta^{(\theta)}_{m,\rho_1,\rho_2}\hbar''|| + \left(\int_x^{\frac{mx+1}{(m+2)}} \frac{\left|\frac{mx+1}{m+2}-u\right|}{\delta^{(\theta)}_{m,\rho_1,\rho_2}(u)}du\right) ||\delta^{(\theta)}_{m,\rho_1,\rho_2}\hbar''||.$$

$$\le \frac{1}{\delta^{(\theta)}_{m,\rho_1,\rho_2}(x)}||\delta^{(\theta)}_{m,\rho_1,\rho_2}\hbar''|| \left[U^{(\theta)}_{m,\rho_1,\rho_2}((t-x)^2;x) + \left(\frac{1-2x}{m+2}\right)^2\right]. \quad (15)$$

Now, using the inequality (9), we get

$$|\overline{U}^{(\theta)}_{m,\rho_1,\rho_2}(\hbar;x) - \hbar(x)| \le \frac{3}{(m+2)}||\delta^{(\theta)}_{m,\rho_1,\rho_2}\hbar''||$$

$$\le \frac{3}{(m+2)}\left(||\varphi^2\hbar''|| + \frac{1}{(m+2)}||\hbar''||\right).$$

Applying (11) and (15), we have for $\mathcal{S} \in C(J)$,

$$|U^{(\theta)}_{m,\rho_1,\rho_2}(\mathcal{S};x) - \mathcal{S}(x)| \le |\overline{U}^{(\theta)}_{m,\rho_1,\rho_2}(\mathcal{S}-\hbar,x)| + |\overline{U}^{(\theta)}_{m,\rho_1,\rho_2}(\hbar;x) - \hbar(x)| + |\hbar(x) - \mathcal{S}(x)|$$

$$+ \left|\mathcal{S}\left(\frac{mx+1}{m+2}\right) - \mathcal{S}(x)\right|$$

$$\le 4||\mathcal{S}-\hbar|| + \frac{3}{(m+2)}||\varphi^2\hbar''|| + \frac{3}{(m+2)^2}||\hbar''|| + \left|\mathcal{S}\left(\frac{mx+1}{m+2}\right) - \mathcal{S}(x)\right|$$

For all $\hbar \in W^2(\varphi)$ using the property of infimum on the right hand side, we have

$$| U_{m,\rho_1,\rho_2}^{(\theta)}(\mathcal{S};x) - \mathcal{S}(x) | \leq 4\tilde{K}_{2,\varphi}\left(\mathcal{S}, \frac{1}{m+2}\right) + \left| \mathcal{S}\left(\frac{mx+1}{m+2}\right) - \mathcal{S}(x) \right|. \tag{16}$$

Also,

$$\left| \mathcal{S}\left(\frac{mx+1}{m+2}\right) - \mathcal{S}(x) \right| = \left| \mathcal{S}\left(x + \frac{1-2x}{m+2}\right) - \mathcal{S}(x) \right|$$
$$\leq \left| \mathcal{S}\left(x + \frac{(1-2x)}{m+2}\right) - \mathcal{S}\left(x - \frac{(1-2x)}{m+2}\right) \right| + \left| \mathcal{S}\left(x - \frac{(1-2x)}{m+2}\right) - \mathcal{S}(x) \right|$$
$$\leq \overrightarrow{\omega_\psi}\left(\mathcal{S},(m+2)^{-1}\right) + \omega\left(\mathcal{S};(m+2)^{-1}\right). \tag{17}$$

Hence, combining (12), (16) and (17), the desired relation is immediate. $\square$

6. Rate of Approximation

In this section, we study the rate of convergence of functions with derivatives of bounded variation.

The class of all absolutely continuous functions $\mathcal{S}$ is denoted by $DBV_{(J)}$, defined and having a derivative $\mathcal{S}'$ on J, analogous to a bounded variation function on J.

The representation of functions $\mathcal{S} \in DBV_{(J)}$ is

$$\mathcal{S}(x) = \int_0^x \hbar(t)dt + \mathcal{S}(0)$$

where $\hbar$ is a bounded variation function on J.

The operators $U_{m,\rho_1,\rho_2}^{(\theta)}(\mathcal{S};x)$ also admit the integral representation

$$U_{m,\rho_1,\rho_2}^{(\theta)}(\mathcal{S};x) = \int_0^1 \mathcal{N}_{m,\rho_1,\rho_2}^{(\theta)}(x,t)\mathcal{S}(t)dt, \tag{18}$$

where the kernel $\mathcal{N}_{m,\rho_1,\rho_2}^{(\theta)}(x,t)$ is given by

$$\mathcal{N}_{m,\rho_1,\rho_2}^{(\theta)}(x,t) = (m+1) \sum_{\mu=0}^{m-\rho_1\rho_2} p_{m-\rho_1\rho_2,\mu}^{(\theta)}(x) \sum_{s=0}^{\rho_2} p_{\rho_2,s}^{(\theta)}(x) p_{m,\mu+s\rho_1}(t).$$

Lemma 5. *For a fixed* $x \in (0,1)$ *and sufficiently large* m, *we have*

(i) $\lambda_{m,\rho_1,\rho_2}^{(\theta)}(x,y) = \int_0^y \mathcal{N}_{m,\rho_1,\rho_2}^{(\theta)}(x,t)dt \leq \dfrac{W_{\rho_1,\rho_2}^{(\theta)}}{(m+2)} \dfrac{x(1-x)}{(x-y)^2}, \; 0 \leq y < x,$

(ii) $1 - \lambda_{m,\rho_1,\rho_2}^{(\theta)}(x,z) = \int_z^1 \mathcal{N}_{m,\rho_1,\rho_2}^{(\theta)}(x,t)dt \leq \dfrac{W_{\rho_1,\rho_2}^{(\theta)}}{(m+2)} \dfrac{x(1-x)}{(z-x)^2}, \; x < z < 1,$

where $W_{\rho_1,\rho_2}^{(\theta)}$ *is given in Lemma* 3.

Proof. (i) From Lemma 3, we get

$$\lambda_{m,\rho_1,\rho_2}^{(\theta)}(x,y) = \int_0^y \mathcal{N}_{m,\rho_1,\rho_2}^{(\theta)}(x,t)dt \leq \int_0^y \left(\frac{x-t}{x-y}\right)^2 \mathcal{N}_{m,\rho_1,\rho_2}^{(\theta)}(x,t)dt$$

$$= U_{m,\rho_1,\rho_2}^{(\theta)}((t-x)^2;x)(x-y)^{-2} \leq \frac{W_{\rho_1,\rho_2}^{(\theta)}}{(m+2)} \frac{x(1-x)}{(x-y)^2}.$$

The (ii) can be proved in the same way hence the details are skipped. $\square$

Theorem 8. *Suppose that $\mathcal{S} \in DBV(\jmath)$. Then for every $x \in (0,1)$ and sufficiently large m, we have*

$$|U_{m,\rho_1,\rho_2}^{(\theta)}(\mathcal{S};x) - \mathcal{S}(x)| \leq \frac{(1-2x)}{(m+2)}\frac{|\mathcal{S}'(x+)+\mathcal{S}'(x-)|}{2} + \sqrt{\frac{\mathcal{W}_{\rho_1,\rho_2}^{(\theta)}x(1-x)}{(m+2)}\frac{|\mathcal{S}'(x+)-\mathcal{S}'(x-)|}{2}}$$

$$+ \frac{\mathcal{W}_{\rho_1,\rho_2}^{(\theta)}(1-x)}{(m+2)}\sum_{s=1}^{[\sqrt{m}]}\bigvee_{x-(x/s)}^{x}(\mathcal{S}_x') + \frac{x}{\sqrt{m}}\bigvee_{x-(x/\sqrt{m})}^{x}(\mathcal{S}_x')$$

$$+ \frac{\mathcal{W}_{\rho_1,\rho_2}^{(\theta)}x}{(m+2)}\sum_{s=1}^{[\sqrt{m}]}\bigvee_{x}^{x+((1-x)/s)}(\mathcal{S}_x') + \frac{(1-x)}{\sqrt{m}}\bigvee_{x}^{x+((1-x)/\sqrt{m})}(\mathcal{S}_x'),$$

where $\bigvee_c^d(\mathcal{S}_x')$ denotes the total variation of $\mathcal{S}_x'$ on $[c,d]$ and $\mathcal{S}_x'$ is defined by

$$\mathcal{S}_x'(t) = \begin{cases} \mathcal{S}'(t) - \mathcal{S}'(x-), & 0 \leq t < x \\ 0, & t = x \\ \mathcal{S}'(t) - \mathcal{S}'(x+) & x < t < 1. \end{cases} \tag{19}$$

Proof. This theorem can be proved in the same way as in ([4], Theorem 7). Hence, the proof of this theorem is skipped. □

7. Numerical Examples

In the following examples, we demonstrate the theoretical results by graphs.

Example 1. *Let $m = 10, \rho_1 = \rho_2 = 1$ and $\theta = 0.5, 0.6, 0.7, 0.8, 1.0$. The convergence of the operators $U_{10,1,1}^{(0.5)}(\mathcal{S};x)$, $U_{10,1,1}^{(0.6)}(\mathcal{S};x)$, $U_{10,1,1}^{(0.7)}(\mathcal{S};x)$, $U_{10,1,1}^{(0.8)}(\mathcal{S};x)$ and $U_{10,1,1}^{(1.0)}(\mathcal{S};x)$ to the function $\mathcal{S}(x) = x^2 e^{\frac{x^3}{x+5}}$ is illustrated in Figure 1.*

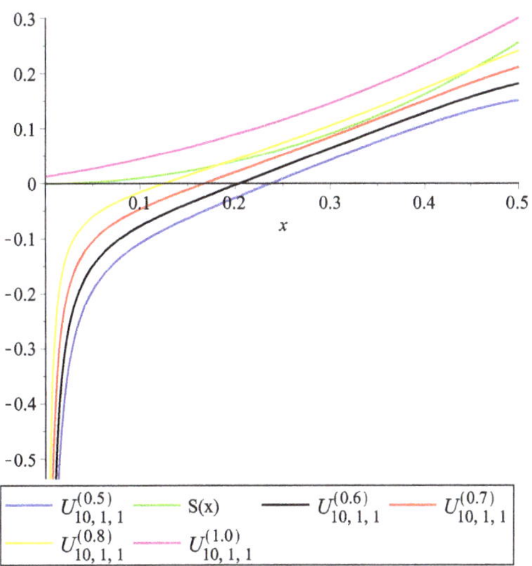

Figure 1. Approximation Process.

Example 2. *Let $m = 50, \rho_1 = \rho_2 = 1$ and $\theta = 0.5, 0.6, 0.7, 0.8, 1.0$. The convergence of the operators $U_{50,1,1}^{(0.5)}(\mathcal{S};x)$, $U_{50,1,1}^{(0.6)}(\mathcal{S};x)$, $U_{50,1,1}^{(0.7)}(\mathcal{S};x)$, $U_{50,1,1}^{(0.8)}(\mathcal{S};x)$ and $U_{50,1,1}^{(1.0)}(\mathcal{S};x)$ to the function $\mathcal{S}(x) = x^3 e^{\frac{x^2}{x+10}}$ is illustrated in Figure 2.*

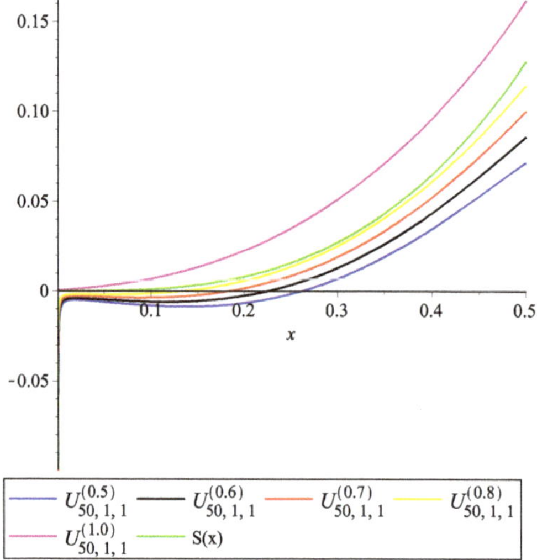

Figure 2. Approximation Process.

8. Conclusions

We have introduced generalized Bernstein–Durrmeyer type operators depending on non-negative integers. We developed many approximation properties such as local and global approximation, the rate of approximation for the Lipschitz type space, Voronovskaja type asymptotic formula and the rate of convergence of functions with derivatives of bounded variation. The constructed operators have better flexibility and rate of convergence which are depending on the selection of the ρ_1, ρ_2 and θ. Graphical representations of our operators for different selections of ρ_1, ρ_2 and θ are also given.

Author Contributions: Project administration, M.M.; Writing—original draft, A.K.; Writing—review and editing, T.A. & M.M. The authors contributed equally and significantly in writing this paper. All authors have read and agreed to the published version of the manuscript.

Funding: The third author has been partially supported within TUBITAK (The Scientific and Technological Research Council of Turkey) 1002-Project 119F191.

Conflicts of Interest: The authors declare no conflict of interest.

References

1. Weierstrass, K. Über die analytische Darstellbarkeit sogennanter willkürlicher Functionen einer reellen Veränderlichen. *Sitzungsber. Akad. Berlin* **1885**, *633–639*, 789–805.
2. Bernstein, S.N. Démonstation du théorème de Weierstrass fondée sur le calcul de probabilités. *Commun. Soc. Math. Kharkow* **1912**, *13*, 1–2.
3. Chen, X.; Tan, J.; Liu, Z.; Xie, J. Approximation of functions by a new family of generalized Bernstein operators. *J. Math. Anal. Appl.* **2017**, *450*, 244–261. [CrossRef]
4. Kajla, A.; Acar, T. Blending type approximation by generalized Bernstein-Durrmeyer type operators. *Miskolc Math. Notes* **2018**, *19*, 319–336. [CrossRef]
5. Acar, T.; Acu, A.M.; Manav, N. Approximation of functions by genuine Bernstein-Durrmeyer type operators. *J. Math. Inequal.* **2018**, *12*, 975–987. [CrossRef]
6. Abel, U.; Heilmann, M. The complete asymptotic expansion for Bernstein-Durrmeyer operators with Jacobi weights. *Mediterr. J. Math.* **2004**, *1*, 487–499. [CrossRef]
7. Cárdenas-Morales, D.; Gupta, V. Two families of Bernstein-Durrmeyer type operators. *Appl. Math. Comput.* **2014**, *248*, 342–353. [CrossRef]

8. Abel, U.; Ivan, M.; Paltanea, R. The Durrmeyer variant of an operator defined by D.D. Stancu. *Appl. Math. Comput.* **2015**, *259*, 116–123.

9. Agrawal, P.N.; Goyal, M.; Kajla, A. $q-$Bernstein-Schurer-Kantorovich type operators. *Boll. Unione Mat. Ital.* **2015**, *8*, 169–180. [CrossRef]

10. Ansari, K.J.; Mursaleen, M.; Rahman, S. Approximation by Jakimovski-Leviatan operators of Durrmeyer type involving multiple Appell polynomials. *Rev. R. Acad. Cienc. Exactas Fis. Nat. Ser. A Mat. RACSAM* **2019**, *113*, 1007–1024. [CrossRef]

11. Acar, T.; Aral, A.; Rasa, I. Positive linear operators preserving τ and τ^2. *Constr. Math. Anal.* **2019**, *2*, 98–102.

12. Mursaleen, M.; Rahman, S.; Ansari, K.J. Approximation by Jakimovski-Leviatan-Stancu-Durrmeyer type operators. *Filomat* **2019**, *33*, 1517–1530. [CrossRef]

13. Acu, A.M.; Kajla, A. *Blending Type Approximation by Generalized Bernstein Operators*; 2020, submitted for publication.

14. Korovkin, P.P. On convergence of linear positive operators in the space of continuous functions. *Dokl. Akad. Nouk. SSR* **1959**, *90*, 961–964. (In Russian)

15. Acar, T.; Aral, A. The new forms of Voronovskaya's theorem in weighted spaces. *Positivity* **2016**, *20*, 25–40. [CrossRef]

16. Acar, T. Quantitative q-Voronovskaya and q-Grüss-Voronovskaya-type results for q-Szasz operators. *Georgian Math. J.* **2016**, *23*, 459–468. [CrossRef]

17. Erencin, A.; Rasa, I. Voronovskaya type theorems in weighted spaces. *Numer. Funct. Anal. Optim.* **2016**, *37*, 1517–1528. [CrossRef]

18. Ulusay, G.; Acar, T. q-Voronovskaya type theorems for q-Baskakov operators. *Math. Method Appl. Sci.* **2016**, *39*, 3391–3401. [CrossRef]

19. Özarslan, M.A.; Aktuğlu, H. Local approximation for certain King type operators. *Filomat* **2013**, *27*, 173–181. [CrossRef]

20. Devore, R.A.; Lorentz, G.G. *Constructive Approximation, Grundlehren der Mathematischen Wissenschaften*; Springer: Berlin, Germany, 1993; Volume 303.

21. Ditzian, Z.; Totik, V. *Moduli of Smoothness*; Springer: New York, NY, USA, 1987.

 © 2020 by the authors. Licensee MDPI, Basel, Switzerland. This article is an open access article distributed under the terms and conditions of the Creative Commons Attribution (CC BY) license (http://creativecommons.org/licenses/by/4.0/).

Article

A Collocation Method Using Radial Polynomials for Solving Partial Differential Equations

Cheng-Yu Ku [1,2] and **Jing-En Xiao [2,*]**

[1]   Center of Excellence for Ocean Engineering, National Taiwan Ocean University, Keelung 20224, Taiwan; chkst26@mail.ntou.edu.tw

[2]   Department of Harbor and River Engineering, National Taiwan Ocean University, Keelung 20224, Taiwan

*   Correspondence: 20452002@email.ntou.edu.tw; Tel.: +886-2-2462-2192 (ext. 6159)

Received: 31 July 2020; Accepted: 23 August 2020; Published: 26 August 2020

Abstract: In this article, a collocation method using radial polynomials (RPs) based on the multiquadric (MQ) radial basis function (RBF) for solving partial differential equations (PDEs) is proposed. The new global RPs include only even order radial terms formulated from the binomial series using the Taylor series expansion of the MQ RBF. Similar to the MQ RBF, the RPs is infinitely smooth and differentiable. The proposed RPs may be regarded as the equivalent expression of the MQ RBF in series form in which no any extra shape parameter is required. Accordingly, the challenging task for finding the optimal shape parameter in the Kansa method is avoided. Several numerical implementations, including problems in two and three dimensions, are conducted to demonstrate the accuracy and robustness of the proposed method. The results depict that the method may find solutions with high accuracy, while the radial polynomial terms is greater than 6. Finally, our method may obtain more accurate results than the Kansa method.

Keywords: multiquadric; radial basis function; radial polynomials; the shape parameter; meshless; Kansa method

1. Introduction

Recently, the meshless approach has raised extensive attention due to its computational efficiency as well as simple collocation scheme. Many varieties of the radial basis functions (RBFs) have been developed for dealing with partial differential equations (PDEs) [1–3]. Most popular RBFs, such as the Gaussian [4–6], multiquadric (MQ) [7,8], and inverse multiquadric (IMQ) [9–11], require the shape parameter. Among them, the Kansa method [12] is recognized as one of the most popular domain-type meshfree approaches for solving PDEs. The MQ RBF adopted by the Kansa method becomes the well-known RBF, which has been successfully adopted for solving numerous engineering problems [13,14]. Despite the success of the Kansa method, limitations regarding to the accuracy affecting by the shape parameter still remain. The MQ RBF depends on the shape parameter that plays an important role for remaining the RBF as a smooth and non-singular function for solving PDEs. Attempts regarding for identifying proper value for the shape parameter of the MQ RBF have been widely studied, such as the LOOCV optimization technique [15–17]. The question of finding the optimal shape parameter in the MQ RBF, however, is still very challenging.

In this study, we propose radial polynomials (RPs) rooted in the MQ RBF for solving PDEs. Formulated from the binomial series using the Taylor series expansion of the MQ RBF, the new global RPs include only even order radial terms. The proposed RPs may be regarded as the equivalent expression of the MQ RBF in series form. Not only are the RPs infinitely smooth and differentiable in nature, but the proposed RPs do not require any extra shape parameters. Therefore, the challenging task for finding the optimal shape parameter in the Kansa method is avoided. Several numerical implementations, including two- and three-dimensional problems, are conducted to verify the accuracy and robustness of the proposed RPs. The structure of this article is organized as follows: In Section 2, formulation of the radial polynomial basis function is presented. To verify the proposed RPs, we conduct a convergence analysis in Section 3. Section 4 is devoted to present several numerical examples in two and three dimensions. The discussion of this paper is addressed in Section 5. Conclusions are given finally.

2. Formulation of the Radial Polynomials

Considering a region, Ω, with the boundary, $\partial\Omega$, the governing equation for the three-dimensional PDE can be expressed as follows.

$$\Delta u(\mathbf{x}) + D\frac{\partial u(\mathbf{x})}{\partial x} + E\frac{\partial u(\mathbf{x})}{\partial y} + F\frac{\partial u(\mathbf{x})}{\partial z} + Gu(\mathbf{x}) = H \text{ in } \Omega, \tag{1}$$

$$u(\mathbf{x}) = g(\mathbf{x}) \text{ on } \partial\Omega^D, \tag{2}$$

$$\frac{\partial u(\mathbf{x})}{\partial n} = f(\mathbf{x}) \text{ on } \partial\Omega^N, \tag{3}$$

in which Δ represents Laplace operator, $\mathbf{x} = (x, y, z)$, $u(\mathbf{x})$ is the unknown, D, E, F, G and H are given functions. Ω is a bounded domain with boundary $\partial\Omega^D$ and $\partial\Omega^N$. $\partial\Omega^D$ denotes boundary subjected to Dirichlet data, $\partial\Omega^N$ denotes boundary subjected to Neumann data, $g(\mathbf{x})$ and $f(\mathbf{x})$ represent given boundary data. The meshless method using the MQ RBF is often named the Kansa method, where the RBFs are directly implemented for the approximation of the solution of partial differential equations. We may express the unknown by the RBF as follows.

$$u(\mathbf{x}) = \sum_{j=1}^{M_c} \lambda_j \varphi(r_j), \tag{4}$$

where r_j is the radial distance, $r_j = |\mathbf{x} - \mathbf{s}_j|$, $\varphi(r_j)$ represents the RBF which is the distance of $\mathbf{x}$ and $\mathbf{s}_j$, $\mathbf{s}_j$ is the center, $\mathbf{x}$ denotes an arbitrary point inside the domain, λ_j is the coefficient to be solved and M_c is the number of the center points. The MQ RBF may be expressed as follows.

$$\varphi(r_j) = \sqrt{r_j^2 + c^2}. \tag{5}$$

With the introduction of the shape parameter, the MQ RBF becomes a smooth and non-singular function. Because the Kansa method is a domain-type method, it has to discretize the governing equation inside the domain using the MQ RBF. We may insert the above equation into Equation (1). After obtaining the MQ RBF derivatives, we may obtain the following equation in two-dimensions.

$$\sum_{j=1}^{M_c} \lambda_j \frac{r_j^2 + 2c^2}{\left(r_j^2 + c^2\right)^{1.5}} + \sum_{j=1}^{M_c} \lambda_j \frac{D\left(x - x_j\right) + E\left(y - y_j\right)}{\left(r_j^2 + c^2\right)^{0.5}} + G\sum_{j=1}^{M_c} \lambda_j \left(r_j^2 + c^2\right)^{0.5} = H \text{ in } \Omega. \tag{6}$$

The above equation demonstrates that the derivatives of the MQ basis function may become singular at the center point ($r_j = 0$) if the shape parameter is zero. It is obvious that the MQ RBF is infinitely differentiable depending on the shape parameter. To avoid the singularity, the shape parameter must not be equal to zero. In this study, we propose RPs based on the MQ RBF without the shape parameter. For the mathematical formulation of the RPs, we may start from the MQ RBF. Equation (5) can be rewritten as follows.

$$\varphi(r_j) = c\sqrt{(r_j/c)^2 + 1}. \tag{7}$$

Using the binomial series from the Taylor series of Equation (7), we have

$$c\sqrt{(r_j/c)^2 + 1} = c\sum_{k=0}^{\infty} \binom{\alpha}{k} \left((r_j/c)^2\right)^k,$$ (8)

where $\binom{\alpha}{k} := \frac{\alpha(\alpha-1)(\alpha-2)\cdots(\alpha-k+1)}{k!}$ and $\alpha = 1/2$.

Using the finite terms, M_n, to approximate the solution, we may express the MQ RBF in series form as follows.

$$\varphi(r_j) = c\sum_{k=0}^{M_n} \binom{0.5}{k}\left(\frac{1}{c}\right)^{2k} r_j^{2k},$$ (9)

where M_n is the order of the radial polynomials. In this study, we propose a novel meshless method to approximate the solution in terms of the RPs as follows.

$$u(\mathbf{x}) = \sum_{j=1}^{M_c} a_j \varphi(r_j),$$ (10)

where M_c represents the center point number. The above equation proves that the MQ RBF can be expressed as a radial polynomial with only even order terms. Equation (8) can be regarded as the equivalent series form of the MQ RBF. Inserting Equation (8) into Equation (10), we have

$$u(\mathbf{x}) \approx \sum_{j=1}^{M_c} a_j c \sum_{k=0}^{M_n} \binom{0.5}{k}\left(\frac{1}{c}\right)^{2k} r_j^{2k}.$$ (11)

Combining the constants in the above equation, we obtain

$$u(\mathbf{x}) \approx \sum_{j=1}^{M_c}\sum_{k=0}^{M_n} b_{j,k} r_j^{2k},$$ (12)

in which $b_{j,k}$ are the coefficients to be solved. Using Equation (12) for the discretization of Equation (1), we may obtain the following equation:

$$\sum_{j=1}^{M_c}\sum_{k=0}^{M_n} b_{j,k} L_{1c} r_j^{2k-2} + \sum_{j=1}^{M_c}\sum_{k=0}^{M_n} b_{j,k} 2k L_{2c} r_j^{2k-2} + G\sum_{j=1}^{M_c}\sum_{k=0}^{M_n} b_{j,k} r_j^{2k} = H \text{ in } \Omega,$$ (13)

where $L_{1c} = 4k^2$, $L_{2c} = (D(x-x_j) + E(y-y_j))$ and $L_{1c} = 4k^2 + 2k$, $L_{2c} = (D(x-x_j) + E(y-y_j) + F(z-z_j))$ are in two and three dimensions, respectively. To determine the unknown coefficients, we apply the approximate solution with the boundary data at collocation points to satisfy the governing equation. We may get the system of simultaneous equations.

$$\mathbf{Ab} = \mathbf{R},$$ (14)

where $\mathbf{b}$ is the unknown coefficient with the size of $N \times 1$ to be evaluated, $\mathbf{R}$ is the known function with the size of $M \times 1$, $\mathbf{A}$ is an $M \times N$ matrix where $M = M_i + M_b$ and $N = M_c \times M_n$. The above equation can be written as follows:

$$\begin{bmatrix} \mathbf{A_I} \\ \mathbf{A_B} \end{bmatrix}[\mathbf{b}] = \begin{bmatrix} \mathbf{R_I} \\ \mathbf{R_B} \end{bmatrix}.$$ (15)

In the preceding equations, $\mathbf{A_I}$ represents the $M_i \times N$ submatrix from the inner collocation points, $\mathbf{A_B}$ represents the $M_b \times N$ submatrix from the boundary collocation points, $\mathbf{R_I}$ is the vector of function values at the inner points which is a $M_i \times 1$ vector, $\mathbf{R}_B$ is the data at the boundary points which is an $M_b \times 1$ vector, M_b is the boundary point number, M_i is the inner point number. The root mean square error (RMSE) is adopted to evaluate the accuracy which is defined by

$$\text{Root mean square error} = \sqrt{\frac{1}{M_m} \sum_{i=1}^{M_m} \left(\hat{u}(x_i) - u(x_i) \right)^2}, \tag{16}$$

in which M_m represents the number of the measuring points with uniform distribution; $u(x_i)$ and $\hat{u}(x_i)$ are the exact and approximate solutions at the ith collocation point, respectively.

3. Accuracy and Convergence Analysis

We first investigate a Laplacian problem in two dimensions enclosed by an irregular domain. The governing equation is

$$\Delta u(\mathbf{x}) = 0, (\mathbf{x}) \in \Omega. \tag{17}$$

The star–like object boundary in two dimensions can be expressed in the following form:

$$\partial \Omega = \left\{ (x,y) \middle| x = \rho(\theta) \cos \theta, y = \rho(\theta) \sin \theta, \rho(\theta) = \sec (3\theta)^{\sin(6\theta)}, 0 \le \theta \le 2\pi \right\}. \tag{18}$$

The exact solution of Equation (17) is designated as

$$u(\mathbf{x}) = e^x \cos(y) + e^y \sin(\mathbf{x}). \tag{19}$$

To verify the accuracy and convergence, we conduct a series of testing cases for the radial polynomial terms in which all cases adopt the same configurations of the boundary, center and inner points as shown in Figure 1. In the analysis, M_b, M_i and M_c are 1208, 151, and 151, respectively. The number of the RPs terms, M_n, needs to be given for the proposed method. As shown in Figure 2, for the RPs, it is found that the RMSE decreases with the increase in the number of RPs terms in which solutions with high accuracy may be found with the radial polynomial terms from 6 to 12.

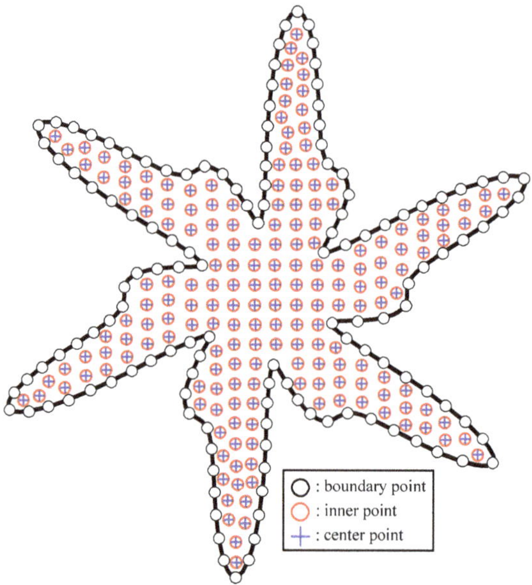

Figure 1. Layout of the collocation points.

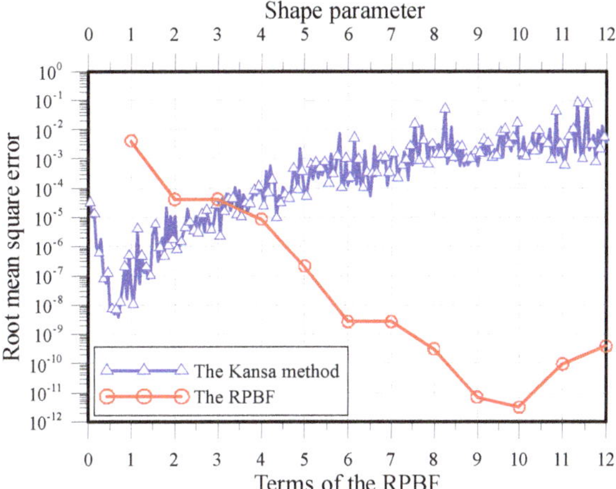

Figure 2. The convergence analysis for the Kansa method and the RPs.

On the other hand, other testing cases using the Kansa method for considering different shape parameters are conducted. Figure 2 shows different shape parameters versus the RMSE. The optimal shape parameter is found within a narrow range of 0.5 to 1. We may also observe that the shape parameter is very sensitive to the accuracy in the Kansa method, such that attempts regarding for identifying proper values for the shape parameter of the Kansa method may be important. It is apparent that the minimums of the RMSE for the Kansa method and the RPs are in the order of 10^{-9} and 10^{-12}, respectively.

In addition, to investigate the accuracy, another convergence analysis for investigating the boundary and inner point number is carried out. Table 1 shows the comparison of this study with the Kansa method. We find that very high accurate results may be obtained using the proposed RPs.

Table 1. Results comparison between this study and the Kansa method with the optimal shape parameter.

			RMSE		Condition Number	
M_b	M_i	M_c	This Study	The Kansa Method (Optimal Shape Parameter)	This Study	The Kansa Method
736	92	92	3.77×10^{-12}	2.96×10^{-9} $(c = 0.70)$	8.33×10^{22}	5.32×10^{20}
1208	151	151	7.29×10^{-12}	2.44×10^{-9} $(c = 0.95)$	1.09×10^{23}	2.63×10^{20}
1792	224	224	7.06×10^{-12}	7.53×10^{-9} $(c = 1.05)$	4.52×10^{23}	7.30×10^{21}
2480	310	310	5.90×10^{-12}	5.70×10^{-9} $(c = 1.05)$	3.86×10^{24}	4.21×10^{20}
3240	405	405	5.31×10^{-12}	9.42×10^{-9} $(c = 1.30)$	1.81×10^{24}	1.52×10^{21}
4115	514	514	4.77×10^{-12}	7.52×10^{-9} $(c = 1.25)$	1.18×10^{24}	2.87×10^{21}

4. Numerical Examples

To investigate the applicability of the proposed RPs, four numerical examples are conducted, in which Sections 4.1 and 4.2 are steady-state linear two-dimensional PDEs, Section 4.3 is the three-dimensional modified Helmholtz equation, and Section 4.4 is the three-dimensional Poisson equation.

4.1. A Two-Dimensional Ameoba-Shaped Problem

We first consider the following two-dimensional PDEs.

$$\Delta u(\mathbf{x}) + D\frac{\partial u(\mathbf{x})}{\partial x} + E\frac{\partial u(\mathbf{x})}{\partial y} + F\frac{\partial u(\mathbf{x})}{\partial z} + Gu(\mathbf{x}) = H, \ \mathbf{x} \in \Omega, \tag{20}$$

where $D = y\cos(x)$, $E = -\sin h(x)$, $F = 0$, $G = x^2 + y^2$. The function, H, can be directly derived from the exact solution as follows:

$$H = \left(-\pi^2 + 1\right)[\sin(\pi x)\cosh(y)] + \left(\pi^2 - 1\right)[\cos(\pi x)\sinh(y)]$$
$$+(\pi y\cos(x) - \sinh(x))[\sin(\pi x)\sinh(y)] + (\pi y\cos(x) + \sinh(x))[\cos(\pi x)\cosh(y)] \qquad (21)$$
$$+\left(x^2 + y^2\right)[\sin(\pi x)\cosh(y) - \cos(\pi x)\sinh(y)]$$

The amoeba-like object boundary in two dimensions is defined as

$$\partial\Omega = \left\{(x, y)\,\Big|\,x - \rho(\theta)\cos\theta, y = \rho(\theta)\sin\theta, \rho(\theta) = e^{(\sin\theta\sin\theta s)^2} + e^{(\cos\theta\cos\theta c)^2}, 0 \le \theta \le 2\pi\right\}. \qquad (22)$$

Both Dirichlet and Neumann boundary conditions are considered as follows:

$$u(\mathbf{x}) = \sin(\pi x)\cosh(y) - \cos(\pi x)\sinh(y), (x, y) \in \partial\Omega^D, \qquad (23)$$

$$\frac{\partial u(\mathbf{x})}{\partial n} = [\nabla(\sin(\pi x)\cosh(y) - \cos(\pi x)\sinh(y))] \cdot \vec{n}, (x, y) \in \partial\Omega^N. \qquad (24)$$

In this example, the Kansa method and the proposed RPs are examined. Figure 3 depicts the configuration of the collocation points. The over-specified Dirichlet as well as Neumann boundary data are imposed on the whole boundary. In the analysis, M_b, M_i and M_c are 1750, 500 and 500, respectively. The analysis of convergence for the RPs terms is conducted, as shown in Figure 4. According to Figure 4, it is found that highly accurate solutions may be solved with the radial polynomial terms from 7 to 12. Consequently, the terms of the RPs are set to 9. The result comparison for the Kansa method and the proposed RPs is shown in Table 2. Table 2 demonstrates that highly accurate results are obtained in which the RMSE of the proposed method is within the order of 10^{-8}. On the other hand, the minimum RMSE for the Kansa method with the optimal shape parameter can only reach to the order of 10^{-4}. Figure 4 demonstrates results of the convergence analysis in which it is found that solutions with high accuracy may be obtained with the radial polynomial terms from 6 to 11. Moreover, it is clear that the number of terms is not very sensitive to the result. Figure 5 depicts the numerical solution is identical to the exact solution.

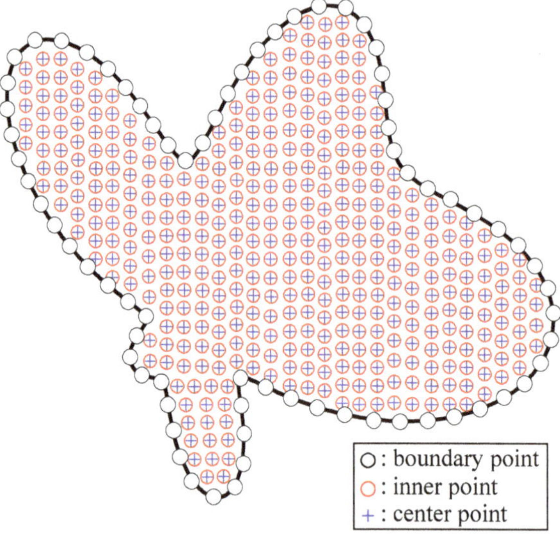

Figure 3. Layout of the collocation points.

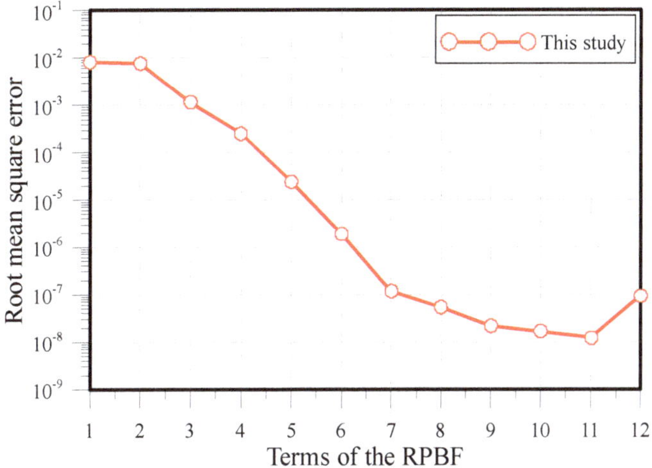

Figure 4. The terms of the RPs versus the RMSE.

Table 2. The RMSE of the RPs and the Kansa method.

M_b	M_i	M_c	RMSE	
			This Study	**The Kansa Method (Optimal Shape Parameter)**
1050	300	300	7.65×10^{-8}	5.46×10^{-4} $(c = 1.00)$
1400	400	400	8.89×10^{-8}	4.72×10^{-4} $(c = 0.95)$
1750	500	500	6.77×10^{-8}	4.14×10^{-4} $(c = 1.05)$
2100	600	600	6.26×10^{-8}	3.79×10^{-4} $(c = 1.05)$
2450	700	700	5.80×10^{-8}	3.56×10^{-4} $(c = 1.30)$
2800	800	800	5.41×10^{-8}	3.45×10^{-4} $(c = 1.25)$

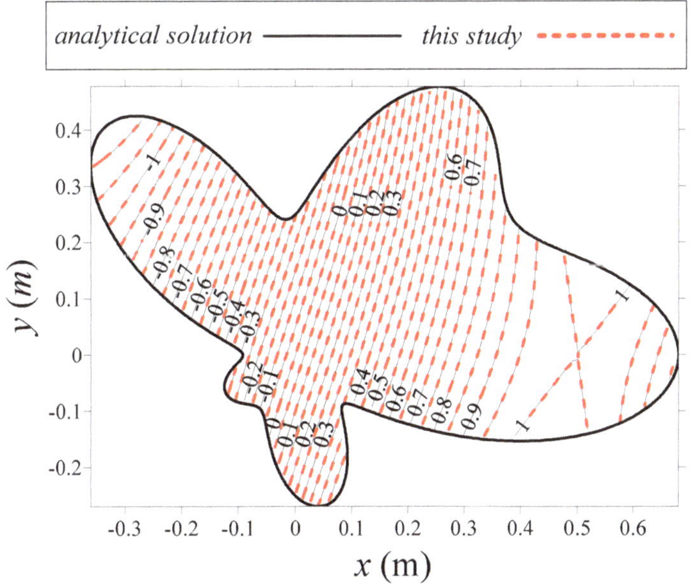

Figure 5. Result comparison between numerical and the analytical solutions.

4.2. A Two-Dimensional Star-Shaped Problem

The second example is a problem enclosed by a star-shaped boundary in two dimensions.

$$\Delta u(\mathbf{x}) + D\frac{\partial u(\mathbf{x})}{\partial x} + E\frac{\partial u(\mathbf{x})}{\partial y} + F\frac{\partial u(\mathbf{x})}{\partial z} + Gu(\mathbf{x}) = H, \; \mathbf{x} \in \Omega, \tag{25}$$

where $D = y^2 \sin(x)$, $E = xe^y$, $F = 0$, $G = \sin(x) + \cos(y)$. The function, H, can be directly derived from the exact solution as follows:

$$H = \left(-\pi^2 y \sin(\pi x) - \pi^2 x \cos(\pi y) \right) + \left(y^2 \sin(x) \right) [\pi y \cos(\pi x) + \cos(\pi y)] \\ + (xe^y) [\sin(\pi x) - \pi x \sin(\pi y)] + (\sin(x) + \cos(y)) [y \sin(\pi x) + x \cos(\pi y)] \tag{26}$$

The star-like object boundary in two dimensions is defined as

$$\partial\Omega = \left\{ (x, y) \big| x = \rho(\theta) \cos\theta, y = \rho(\theta) \sin\theta, \rho(\theta) = 1 + (\cos 4\theta)^2, 0 \le \theta \le 2\pi \right\}. \tag{27}$$

The Dirichlet boundary conditions are considered as follows:

$$u(\mathbf{x}) = y \sin(\pi x) + x \cos(\pi y), (x, y) \in \partial\Omega^D. \tag{28}$$

In this example, the Dirichlet data are applied on the whole boundary using Equation (28). In the analysis, M_b, M_i and M_c are 1800, 200 and 200, respectively. We conduct the convergence analysis for the RPs terms. Figure 6 displays the configuration of the boundary, inner and center collocation points. Figure 7 displays the terms of the RPs versus the RMSE in which we may find that solutions with high accuracy in the order of 10^{-8} may be found with the radial polynomial terms from 7 to 12. Consequently, the terms of the RPs are set to 9.

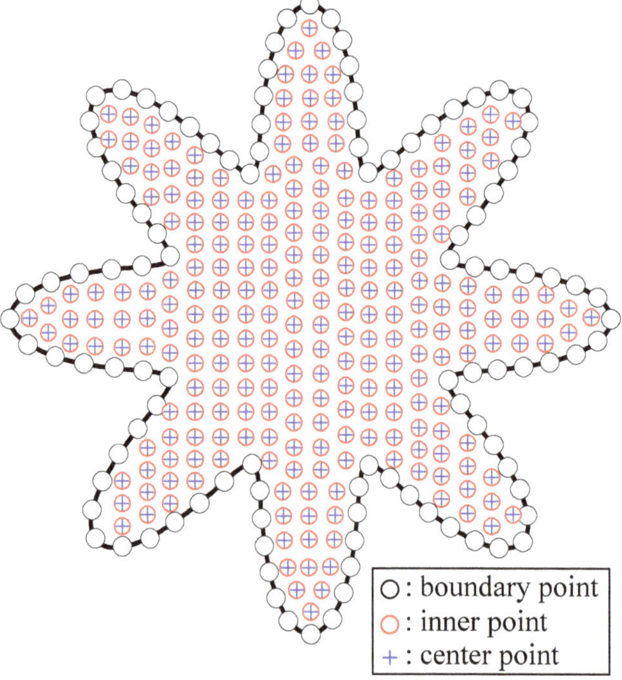

Figure 6. Layout of the collocation points.

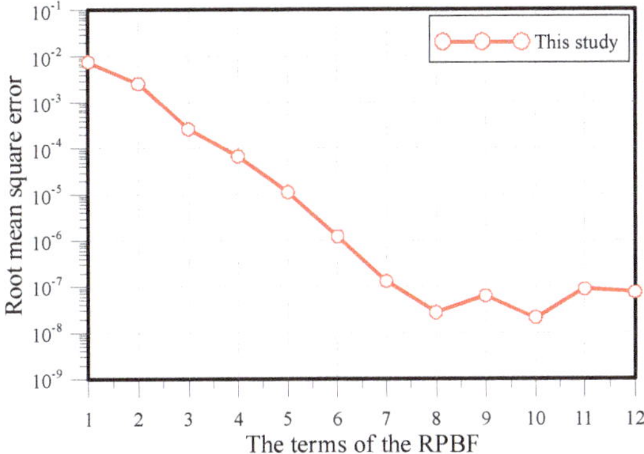

Figure 7. The terms of the RPs versus the RMSE.

Figures 8 and 9 demonstrate the RMSE versus the boundary and inner point numbers, respectively. We may find that promising solutions may be found while the boundary and inner point numbers are greater than 500 and 100, respectively. Figure 10 demonstrates the comparison of the analytical and the numerical solutions. It can be found that the results agree with the analytical solutions.

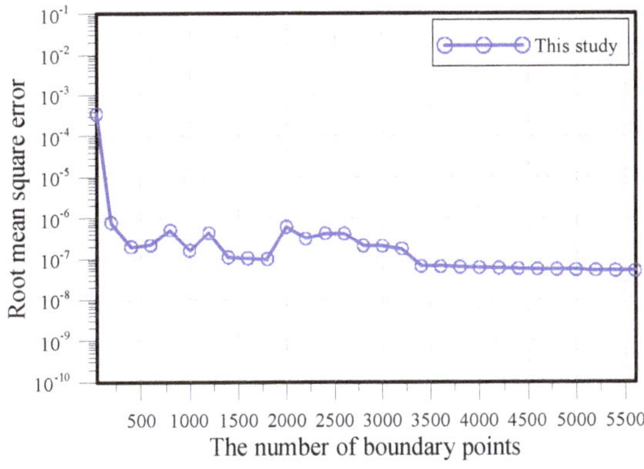

Figure 8. The convergence analysis for the boundary point number.

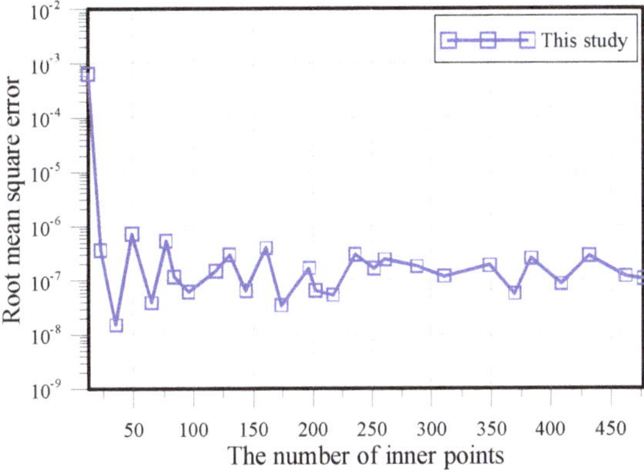

Figure 9. The convergence analysis for the inner point number.

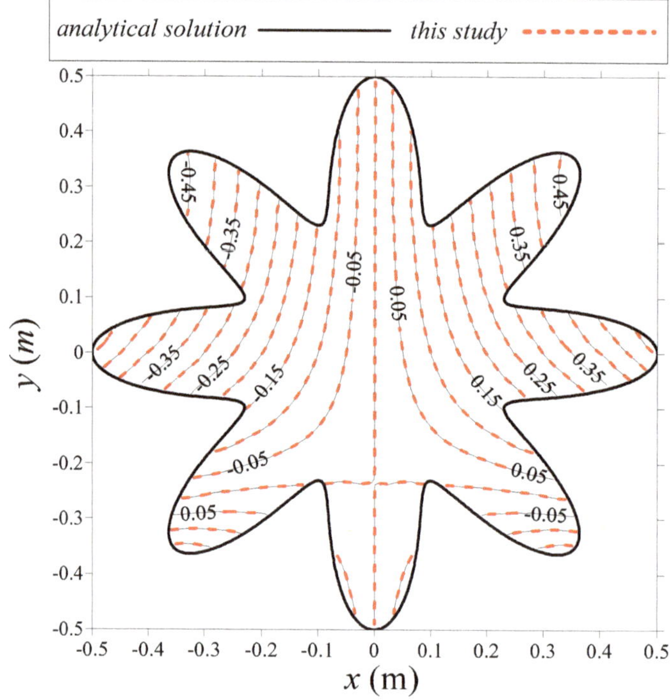

Figure 10. Result comparison between numerical and the analytical solutions.

4.3. A Three-Dimensional Modified Helmholtz Problem

Consider a three-dimensional modified Helmholtz equation. The equation is written as follows.

$$\Delta u(\mathbf{x}) + Gu(\mathbf{x}) = H, \ \mathbf{x} \in \Omega, \tag{29}$$

where $D = E = F = 0$, $G = -\lambda^2$, $H = (1 - \lambda^2)(e^x + e^y + e^z) + xyz$, λ represents wave number and $\lambda = 100$. The domain in three dimensions can be expressed in the following form.

$$\partial\Omega = \left\{ (x, y, z) \big| x = \rho(\theta)\sin\theta\cos\varphi, y = \rho(\theta)\sin\theta\sin\varphi, z = \rho(\theta)\cos\varphi \right\}, \tag{30}$$

where $\rho(\theta, \varphi) = 1 + 1/8\sin(10\,\theta)\sin(9\,\varphi), 0 \le \theta \le 2, 0 \le \varphi \le \pi$. The Dirichlet boundary data are applied on $\partial\Omega$ using the following exact solution.

$$u(\mathbf{x}) = e^x + e^y + e^z - xyz/\lambda^2, (x, y) \in \partial\Omega^D. \tag{31}$$

In this example, the layout of the domain is depicted in Figure 11. The Dirichlet data are applied on the whole boundary using Equation (31). In the analysis, M_b, M_i and M_c are 7569, 800 and 800, respectively. Figure 12 demonstrates solutions with high accuracy in the order of 10^{-8} may be found with the radial polynomial terms from 8 to 11. Consequently, the terms of the RPs are set to 9. The result comparison for the Kansa method and the proposed RPs is shown in Table 3. Table 3 demonstrates that highly accurate results are obtained in which the RMSE is within the order of 10^{-11}. On the other hand, the best RMSE for the Kansa method with the optimal shape parameter can only reach to the order of 10^{-7}.

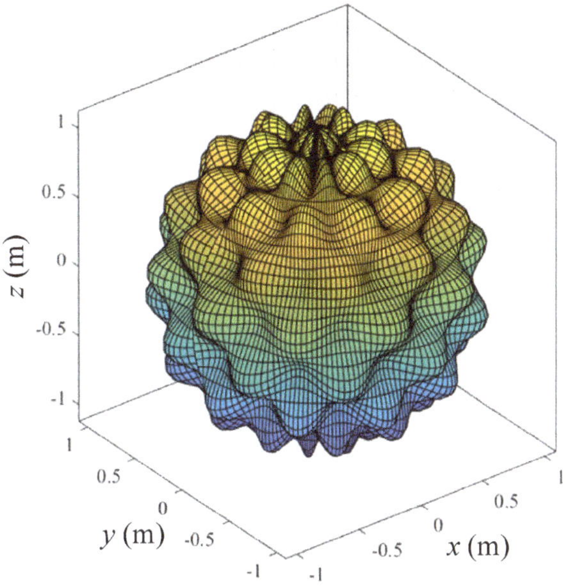

Figure 11. Layout of the three-dimensional modified Helmholtz equation.

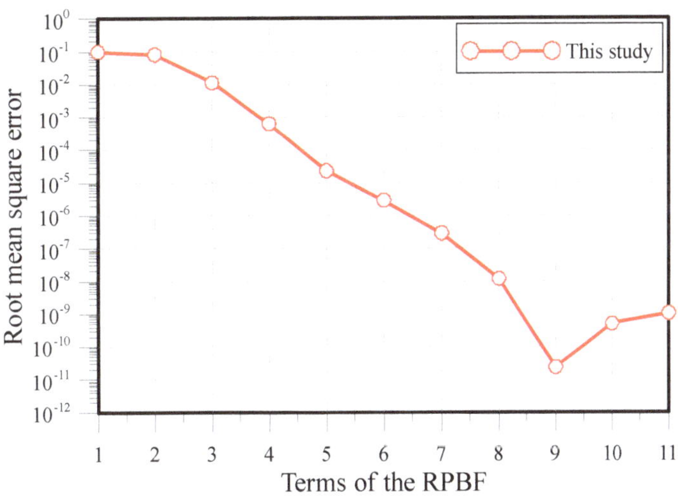

Figure 12. The terms of the RPs versus the RMSE.

Table 3. The RMSE of the RPs and the Kansa method.

M_b	M_i	M_c	RMSE	
			This Study	The Kansa Method (Optimal Shape Parameter)
6724	700	700	4.28×10^{-11}	1.10×10^{-6} ($c = 1.30$)
7569	800	800	2.32×10^{-11}	8.48×10^{-7} ($c = 1.30$)
8100	900	900	4.40×10^{-11}	1.01×10^{-6} ($c = 1.90$)
9025	1000	1000	5.99×10^{-11}	1.08×10^{-6} ($c = 1.10$)
10,000	1100	1100	5.03×10^{-11}	7.88×10^{-7} ($c = 1.30$)

4.4. A Three-Dimensional Poisson Problem

The last example under consideration is a three-dimensional Poisson equation enclosed by an irregular domain. The governing equation is expressed as follows.

$$\Delta u(\mathbf{x}) = H, \ \mathbf{x} \in \Omega, \tag{32}$$

where $D = E = F = G = 0$ and $H = -\sin x - \cos y - \sin z$. The domain in three dimensions can be expressed in the following parametric equation.

$$\partial \Omega = \left\{ (x, y, z) \big| x = \rho(\theta) \cos \theta, y = \rho(\theta) \sin \theta \sin \varphi, z = \rho(\theta) \sin \theta \cos \varphi \right\}, \tag{33}$$

where $\rho(\theta) = \left[\cos(3\theta) + \sqrt{8 - \sin^2(3\theta)} \right]^{1/3}$. The Dirichlet boundary data are assigned on $\partial \Omega^D$ using the following exact solution.

$$u(\mathbf{x}) = \sin(x) + \cos(y) + \sin(z), (x, y) \in \partial \Omega^D. \tag{34}$$

The layout of the domain is depicted in Figure 13. The Dirichlet boundary conditions are given on the irregular domain in three dimensions using Equation (34). In the analysis, M_b, M_i and M_c are 7357, 756 and 756, respectively. Figure 14 shows the RMSE versus the terms of the RPs. It is apparent that the promising numerical solution in the order of 10^{-8} may be obtained while the M_n is greater than 6. Consequently, the terms of the RPs is set to 9. Additionally, several cases for evaluating the number of the collocation points to the accuracy are conducted in Table 4. According to Table 4, it depicts that the accuracy can reach up to the order of 10^{-10}.

Table 4. The RMSE of the RPs.

M_b	M_i	M_c	RMSE This Study
5706	630	630	1.75×10^{-10}
7357	756	756	1.82×10^{-10}
9208	882	882	1.87×10^{-10}
11,259	1008	1008	1.92×10^{-10}
13,510	1134	1134	1.94×10^{-10}

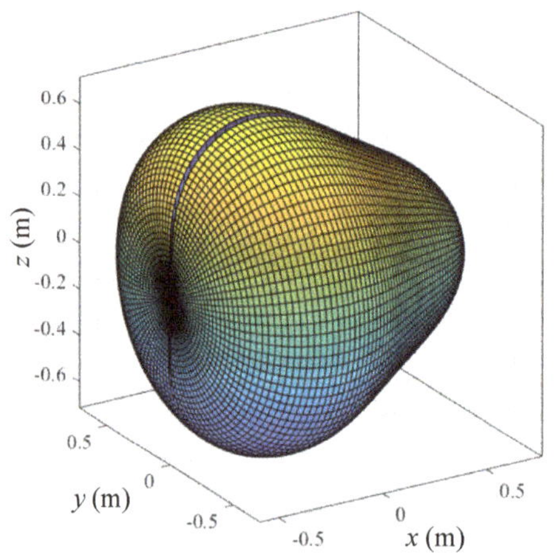

Figure 13. Layout of the three-dimensional Poisson equation.

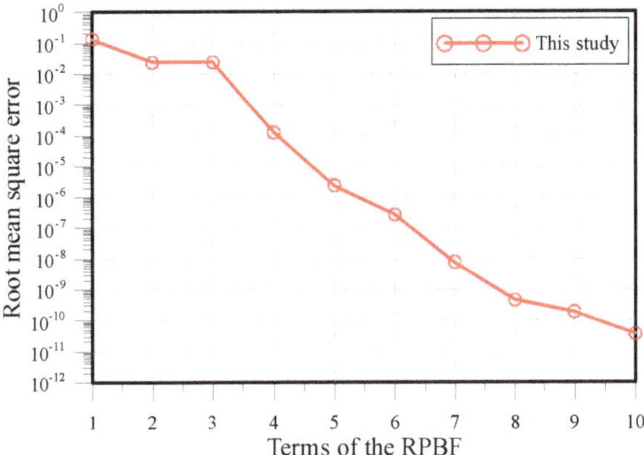

Figure 14. The terms of the RPs versus the RMSE.

5. Discussion

This study presents a collocation method using RPs which is regarded as the equivalent expression of the MQ RBF in series form for PDEs. The conception of the new global RPs includes only even order radial terms formulated from the binomial series using the Taylor series expansion of the MQ RBF. The discussions for this study are as follows.

The MQ RBF adopted by the Kansa method becomes one of the most successful RBFs for solving numerous problems. With the introduction of the shape parameter, the MQ RBF becomes a smooth and non-singular function. Even though the MQ RBF and its derivatives are smooth and global infinitely differentiable, the discretization of the governing equation may become singular while $c = 0$ at $r_j = 0$. It is obvious the shape parameter plays a role for shifting from the singularity while the center point is coincided with the inner point. However, the near singular effects still remain. This may explain that the accuracy in the Kansa method is strongly affected by the shape parameter.

To deal with the issue, we adopt the RPs as the basis function in which the proposed RPs are the equivalent expression of the MQ RBF in series form. It is advantageous that the proposed RPs and their derivatives are infinitely smooth and differentiable in nature without using the shape parameter. Because RPs are a non-singular series function, there are no near singular or singular effects at all. Accordingly, the method may obtain more accurate solutions than the Kansa method in our numerical implementations. In addition, accurate results can be directly obtained without using the tedious procedure for finding the optimal shape parameter.

Even though the shape parameter is not required in the proposed method, the radial polynomial terms have to be decided in advance. From the numerical implementations, solutions with high accuracy in the order of 10^{-8} may be found with radial polynomial terms from 6 to 12. The radial polynomial terms are selected to be 9 in our numerical examples. It demonstrates that the radial polynomial terms are considerably less significant to the accuracy than the shape parameter. In the numerical examples, it is found that satisfactory solutions could be obtained while the terms of the RPs are within the range of 6 to 12.

6. Conclusions

A mathematical formulation of the RPs from the binomial series using the Taylor series expansion of the MQ RBF is presented. We prove that the proposed RPs are an equivalent expression of the MQ RBF in series form. Highly accurate results can be directly obtained without using the tedious procedure for finding the optimal shape parameter. Additionally, numerical comparisons reveal that the presented RPs could obtain better accurate solutions than those of the MQ RBF, even with the optimal shape parameter for solving multi-dimensional PDEs.

Author Contributions: Writing—review and editing, J.-E.X. and C.-Y.K.; conceptualization, C.-Y.K.; methodology, C.-Y.K. All authors have read and agreed to the published version of the manuscript.

Funding: This research was partially supported by the Ministry of Science and Technology of R.O.C.

Acknowledgments: The authors would like to thank the referees for invaluable comments.

Conflicts of Interest: The authors declare no conflict of interest.

References

1. Yao, G.; Tsai, C.H.; Chen, W. The comparison of three meshless methods using radial basis functions for solving fourth-order partial differential equations. *Eng. Anal. Bound. Elem.* **2010**, *34*, 625–631. [CrossRef]
2. Uddin, M.; Haq, S. RBFs approximation method for time fractional partial differential equations. *Commun. Nonlinear Sci.* **2011**, *16*, 4208–4214. [CrossRef]
3. Yao, G.; Chen, C.S.; Zheng, H. A modified method of approximate particular solutions for solving linear and nonlinear PDEs. *Numer. Meth. Part. Differ. Equ.* **2017**, *33*, 1839–1858. [CrossRef]
4. Alipanah, A.; Esmaeili, S. Numerical solution of the two-dimensional Fredholm integral equations using Gaussian radial basis function. *J. Comput. Appl. Math.* **2011**, *235*, 5342–5347. [CrossRef]
5. Lamichhane, A.R.; Chen, C.S. The closed-form particular solutions for Laplace and biharmonic operators using a Gaussian function. *Appl. Math. Lett.* **2015**, *46*, 50–56. [CrossRef]
6. Karimi, N.; Kazem, S.; Ahmadian, D.; Adibi, H.; Ballestra, L.V. On a generalized Gaussian radial basis function: Analysis and applications. *Eng. Anal. Bound. Elem.* **2020**, *112*, 46–57. [CrossRef]
7. Reutskiy, S.Y. Meshless radial basis function method for 2D steady-state heat conduction problems in anisotropic and inhomogeneous media. *Eng. Anal. Bound. Elem.* **2016**, *66*, 1–11. [CrossRef]
8. Sarra, S.A. Integrated multiquadric radial basis function approximation methods. *Comput. Math. Appl.* **2006**, *51*, 1283–1296. [CrossRef]
9. Soleymani, F.; Barfeie, M.; Haghani, F.K. Inverse multi-quadric RBF for computing the weights of FD method: Application to American options. *Commun. Nonlinear Sci.* **2018**, *64*, 74–88. [CrossRef]
10. Sivaram, S.A.; Vinoy, K.J. Inverse multiquadric radial basis functions in eigenvalue analysis of a circular waveguide using radial point interpolation method. *IEEE Microw. Wirel. Co.* **2020**, *30*, 537–540. [CrossRef]
11. Kamath, A.; Manzhos, S. Inverse Multiquadratic Functions as the Basis for the Rectangular Collocation Method to Solve the Vibrational Schrödinger Equation. *Mathematics* **2018**, *6*, 253. [CrossRef]
12. Kansa, E.J.; Hon, Y.C. Circumventing the ill-conditioning problem with multiquadric radial basis functions: Applications to elliptic partial differential equations. *Comput. Math. Appl.* **2000**, *39*, 123–137. [CrossRef]
13. Hashemi, M.R.; Hatam, F. Unsteady seepage analysis using local radial basis function-based differential quadrature method. *Appl. Math. Model.* **2011**, *35*, 4934–4950. [CrossRef]
14. Xie, H.; Li, D. A meshless method for Burgers' equation using MQ-RBF and high-order temporal approximation. *Appl. Math. Model.* **2013**, *37*, 9215–9222.
15. Zheng, H.; Yao, G.; Kuo, L.H.; Li, X.X. On the Selection of a Good Shape Parameter of the Localized Method of Approximated Particular Solutions. *Adv. Appl. Math. Mech.* **2018**, *10*, 896–911. [CrossRef]
16. Rippa, S. An algorithm for selecting a good value for the parameter c in radial basis function interpolation. *Adv. Comput. Math.* **1999**, *11*, 193–210. [CrossRef]
17. Cavoretto, R.; De Rossi, A.; Mukhametzhanov, M.S.; Sergeyev, Y.D. On the search of the shape parameter in radial basis functions using univariate global optimization methods. *J. Glob. Optim.* **2019**, *2019*, 1–23. [CrossRef]

 © 2020 by the authors. Licensee MDPI, Basel, Switzerland. This article is an open access article distributed under the terms and conditions of the Creative Commons Attribution (CC BY) license (http://creativecommons.org/licenses/by/4.0/).

 *symmetry*

Article

On the Decomposability of the Linear Combinations of Euler Polynomials with Odd Degrees

Ákos Pintér [1,*] and Csaba Rakaczki [2]

[1] Institute of Mathematics, MTA-DE Research Group "Equations, Functions and Curves", Hungarian Academy of Sciences and University of Debrecen, P. O. Box 12, H-4010 Debrecen, Hungary

[2] Institute of Mathematics, University of Miskolc, H-3515 Miskolc Campus, Hungary; matrcs@uni-miskolc.hu

* Correspondence: apinter@science.unideb.hu

Received: 2 April 2019; Accepted: 18 May 2019; Published: 31 May 2019

Abstract: In the present paper we prove that under certain conditions the linear combination of two Euler polynomials with odd degrees $P_{n,m}(x) = E_n(x) + cE_m(x)$ is always indecomposable over $\mathbb{C}$, where c denotes a rational number.

Keywords: Euler polynomials; higher degree equations

MSC: 11B68; 11B83; 11R09

1. Introduction

Let $\mathbb{K}$ be a field. If $f(x) \in \mathbb{K}[x]$ has degree at least 2, we say that $f(x)$ is *decomposable* over the field $\mathbb{K}$ if we can write $f(x) = f_1(f_2(x))$ for some nonlinear polynomial $f_1(x), f_2(x) \in \mathbb{K}[x]$. Otherwise, we say that $f(x)$ is *indecomposable* over $\mathbb{K}$. Two decompositions $f(x) = f_1(f_2(x))$ and $f(x) = F_1(F_2(x))$ are said to be *equivalent* over the field $\mathbb{K}$, written $f_1 \circ f_2 \sim_{\mathbb{K}} F_1 \circ F_2$, if there exists a linear polynomial $l(x) \in \mathbb{K}[x]$ such that

$$f_1(x) = F_1(l(x)) \quad \text{and} \quad F_2(x) = l(f_2(x)).$$

For a given $f(x) \in \mathbb{K}[x]$ with degree at least 2, a *complete decomposition* of $f(x)$ over $\mathbb{K}$ is a decomposition $f = f_1 \circ \cdots \circ f_k$, where the polynomials $f_i \in \mathbb{K}[x]$ are indecomposable over $\mathbb{K}$ for $i = 1, \ldots, k$. A polynomial of degree greater than 1 always has a complete decomposition, but it does not need to be unique even up to equivalence.

Euler polynomials are defined by the following generating function

$$\sum_{k=0}^{\infty} E_k(x) \frac{t^k}{k!} = \frac{2e^{tx}}{(e^t + 1)}.$$

These polynomials play a central role in various branches of mathematics; for example, in various approximation and expansion formulas in discrete mathematics and in number theory (see for instance [1,2]), in p-adic analyis (see [3], Chapter 2), in statistical physics as well as in semi-classical approximations to quantum probability distributions (see [4–7]).

There are several results connected to the decomposability of an infinite family of polynomials, see for instance [8–12]. Bilu, Brindza, Kirschenhofer, Pintér and Tichy [13] gave all the decompositions of Bernoulli polynomials. Kreso and Rakaczki [14] characterized the all possible decomposations of Euler polynomials with degree even, moreover they showed that every Euler polynomial with odd degree is indecomposable. It is harder to obtain similar results for the sum of polynomials. Pintér and Rakaczki [15] describe the complete decomposition of linear combinations of the form

$$R_n(x) = B_n(x) + cB_{n-2}(x)$$

of Bernoulli polynomials, where c is an arbitrary rational number. Later, Pintér and Rakaczki in [16] proved that for all odd $n > 1$ integer and for all rational number c the polynomials $E_n(x) + cE_{n-2}(x)$ are indecomposable.

The main purpose of this paper is to prove that under certain conditions a linear combination with rational coefficients of two Euler polynomials with odd degrees is always indecomposable. We have

Theorem 1. *Let $P_{n,m}(x) = E_n(x) + cE_m(x)$, where $c = A/B$ is an arbitrary rational number, where $B \neq 2^a$, $a \in \mathbb{N} \cup \{0\}$, n, m are odd integers with $n > m > n/3$. Then the polynomials $P_{n,m}(x)$ are indecomposable over $\mathbb{C}$.*

2. Auxiliary Results

In the first lemma we collect some well known properties of the Euler polynomials which will be used in the sequel, sometimes without particular reference.

Lemma 1. *(a)* $\quad E_n(x) = (-1)^n E_n(1-x);$
(b) $\quad E_n(x+1) + E_n(x) = 2x^n;$
(c) $\quad E'_n(x) = nE_{n-1}(x);$
(d) $\quad E_{2n-1}(1/2) = E_{2n}(0) = E_{2n}(1) = 0 \text{ for } n \in \mathbb{N};$
(e) $\quad E_n(x) = \sum_{k=0}^{n} \binom{n}{k} E_k(0) x^{n-k};$

Proof. See [2]. $\square$

The following result is a general theorem from the theory of decomposability.

Lemma 2 (Kreso and Rakaczki [14]). *Let $F(x) \in \mathbb{K}[x]$ be a monic polynomial such that $\deg F$ is not divisible by the characteristic of the field $\mathbb{K}$. Then for every nontrivial decomposition $F = F_1 \circ F_2$ over any field extension $\mathbb{L}$ of $\mathbb{K}$, there exists a decomposition $F = \tilde{F}_1 \circ \tilde{F}_2$ such that the following conditions are satisfied*

- *$\tilde{F}_1 \circ \tilde{F}_2$ and $F_1 \circ F_2$ are equivalent over $\mathbb{L}$,*
- *$\tilde{F}_1(x)$ and $\tilde{F}_2(x)$ are monic polynomials with coefficients in $\mathbb{K}$,*
- *$\mathrm{coeff}\left(x^{\deg \tilde{F}_1 - 1}, \tilde{F}_1(x)\right) = 0.$*

Moreover, such decomposition $\tilde{F}_1 \circ \tilde{F}_2$ is unique.

Lemma 3. *Let $h(x) \in \mathbb{Q}[x]$ with $\deg h(x) \geq 4$. If $h(x)$ is decomposable over $\mathbb{Q}$ then we can write the polynomial $h(x)$ in the form $h(x) = \frac{u}{v} f(g(x))$, where u and $v \neq 0$ are relative prime integers, $f(x)$ and $g(x) \in \mathbb{Z}[x]$ are primitive polynomials. Moreover, if $h(x)$ is a monic polynomial, then $u = 1$.*

Proof. Suppose that $h(x) = F(G(x))$, where $F(x), G(x) \in \mathbb{Q}[x]$. Let

$$F(x) = b_k x^k + b_{k-1} x^{k-1} + \cdots + b_1 x + b_0,$$
$$G(x) = c_t x^t + c_{t-1} x^{t-1} + \cdots + c_1 x + c_0.$$

Every polynomial with rational coefficients can be written uniquely as a product of a rational number and a primitive polynomial. Hence, we can assume that

$$G(x) = \frac{c}{d} g(x), \text{ where } g(x) \text{ is a primitive polynomial}, c, d \neq 0 \in \mathbb{Z}$$

and so

$$F(G(x)) = b_k \left(\frac{c}{d}\right)^k g(x)^k + b_{k-1} \left(\frac{c}{d}\right)^{k-1} g(x)^{k-1} + \cdots + b_1 \frac{c}{d} g(x) + b_0.$$

The polynomial

$$F_1(x) = b_k \left(\frac{c}{d}\right)^k x^k + b_{k-1} \left(\frac{c}{d}\right)^{k-1} x^{k-1} + \cdots + b_1 \frac{c}{d} x + b_0 \in \mathbb{Q}[x]$$

can be written in the from $\frac{u}{v} f(x)$, where $f(x) \in \mathbb{Z}[x]$ is a primitive polynomial, $u > 0$, $v \neq 0$ are relative prime integers. However, then we have

$$h(x) = F(G(x)) = F_1(g(x)) = \frac{u}{v} f(g(x)). \tag{1}$$

If the polynomial $h(x)$ is monic, then comparing the leading coefficients in (1) one can deduce that $v = u f_k g_t^k$, where f_k and g_t denotes the leading coefficient of the polynomial $f(x)$ and $g(x)$, respectively. This means that u divides v that is $u = 1$. $\square$

Let

$$S^+ = \{f(x) \in \mathbb{C}[x] \mid \; f(x) = f(1-x)\}$$

and

$$S^- = \{f(x) \in \mathbb{C}[x] \mid \; f(x) = -f(1-x)\}.$$

From these definitions it is easy to see that S^+ and S^- are subspaces in the vector space $\mathbb{C}[x]$.

Lemma 4. *Let $P(x) \in \mathbb{Q}[x]$ be a monic polynomial. Assume that $P(x) \in S^-$ and $P(x) = f(g(x))$, where $f(x)$, $g(x) \in \mathbb{Q}[x]$ and $\deg(f(x)), \deg(g(x)) > 1$. Then we can assume that $f(x)$, $g(x)$ are monic, $g(x) \in S^-$ and $f(x) = -f(-x)$.*

Proof. See [16]. $\square$

The following Lemma is a simple combination of Lemmas 3 and 4.

Lemma 5. *Let $P(x) \in \mathbb{Q}[x]$ be a monic polynomial. Assume that $P(x) \in S^-$ and $P(x) = F(G(x))$, where $F(x)$, $G(x) \in \mathbb{Q}[x]$ and $\deg(F(x)) > 1$, $\deg(G(x)) > 1$. Then we can assume that $P(x) = \frac{1}{v} f(g(x))$, where $v \neq 0$ is an integer, $f(x)$ and $g(x)$ are primitive polynomials, $g(x) \in S^-$ and $f(x) = -f(-x)$.*

Proof. From Lemma 4 we can assume that $G(x) \in S^-$ and $F(x) = -F(-x)$. Using the proof of Lemma 3 and the fact that S^- is a subspace of $\mathbb{C}[x]$ we get the assertion of our Lemma. $\square$

Lemma 6. *Let $g(x) = c_t x^t + c_{t-1} x^{t-1} + \cdots + c_1 x + c_0 \in S^-$. Then*

$$- 2c_s = \binom{s+1}{s} c_{s+1} + \binom{s+2}{s} c_{s+2} + \cdots + \binom{t-1}{s} c_{t-1} + \binom{t}{s} c_t \tag{2}$$

for even index $0 \leq s \leq t - 1$.

Proof. Since $g(x) \in S^-$ we have that $-g(x) = g(1-x)$. Computing the coefficient of x^s on the both sides we obtain that

$$-c_s = (-1)^s \left(c_s + \binom{s+1}{s} c_{s+1} + \binom{s+2}{s} c_{s+2} + \cdots + \binom{t-1}{s} c_{t-1} + \binom{t}{s} c_t \right).$$

$\square$

Lemma 7. *Let*

$$f(x) = b_k x^k + b_{k-2} x^{k-2} + b_{k-3} x^{k-3} + \cdots + b_1 x + b_0,$$

$$g(x) = c_t x^t + c_{t-1} x^{t-1} + \cdots + c_1 x + c_0 \in \mathbb{Q}[x].$$

If $k, t \geq 2$ then the coefficient of the monomial x^{kt-2} in the polynomial $f(g(x))$ is

$$b_k \left(k c_t^{k-1} c_{t-2} + \binom{k}{2} c_t^{k-2} c_{t-1}^2 \right).$$

Proof. It is easy to see that the monomial x^{kt-2} occurs only in the term $b_k g(x)^k$. Expanding $g(x)^k$ we simply get the assertion. $\square$

3. Proof of the Theorem

Let n, m be odd positive integers with $n - 2 > m > n/3$, B is an arbitrary integer which is not a power of two. The case of $m = n - 2$ was treated in [16]. Suppose that $P_{n,m}(x)$ is decomposable over $\mathbb{C}$. From Lemmas 2 and 5 we can assume that $P_{n,m}(x) = \frac{1}{v} f(g(x))$, where $v \neq 0$ is an integer, $f(x)$, $g(x) \in \mathbb{Z}[x]$ are primitive polynomials and $g(x) \in S^-$, $f(x) = -f(-x)$. Let

$$f(x) = b_k x^k + b_{k-2} x^{k-2} + b_{k-4} x^{k-4} + \cdots + b_3 x^3 + b_1 x,$$

$$g(x) = c_t x^t + c_{t-1} x^{t-1} + \cdots + c_1 x + c_0.$$

Using (b) of Lemma 1 one can deduce that

$$\frac{1}{v} f(g(x+1)) + \frac{1}{v} f(g(x)) = P_{n,m}(x+1) + P_{n,m}(x) = 2x^n + 2cx^m. \tag{3}$$

Since $P_{n,m}(x) \in S^-$ thus $P_{n,m}(x+1) = -P_{n,m}(-x)$. From (3) we infer that the polynomial $g(x) - g(-x)$ divides the polynomial $P_{n,m}(x) - P_{n,m}(-x) = 2x^n + 2cx^m$, that is

$$g(x) - g(-x) = dx^s h(x), \tag{4}$$

where $d \in \mathbb{Q}, 0 \leq s \leq m$ and the polynomial $h(x)$ divides the polynomial $x^{n-m} + c$ in $\mathbb{Q}[x]$. We know that

$$g(x) - g(-x) = 2c_t x^t + 2c_{t-2} x^{t-2} + \cdots + 2c_3 x^3 + 2c_1 x. \tag{5}$$

If the polynomial $h(x)$ is a constant polynomial then we have $t = s$ and so $c_{t-2} = 0$. It follows from $P_{n,m}(x) = E_n(x) + cE_m(x)$ and (d), (e) of Lemma 1 that the coefficient of x^{n-2} in $P_{n,m}(x)$ equals 0. Applying now Lemma 7 we get that

$$b_k \binom{k}{2} c_t^{k-2} c_{t-1}^2 = 0,$$

which is impossible since $-2c_{t-1} = tc_t$ by Lemma 6.

In the case when $h(x) = x^{n-m} + c$ we get $s + n - m = t$, $d = 2c_t$ and $g(x) - g(-x) = 2c_t x^{s+n-m} + 2c_t cx^s = 2c_t x^t + 2c_t cx^{t-(n-m)}$. Since by assumption $n - m > 2$, we obtain again that $c_{t-2} = 0$, which is not possible.

Next suppose that $1 \leq \deg h(x) < n - m$. In this case one can deduce that s is odd and $h(x) = h(-x)$. Consider first when $1 < s$. Then $c_1 = c_3 = \cdots = c_{s-2} = 0$ and $c_s \neq 0$. Let $G(x) = g(x) - g(0)$ and $F(x) = f(x + g(0))$. Then $f(g(x)) = F(G(x))$, $G(0) = 0$ and

$$G(x) - G(-x) = g(x) - g(-x) = 2c_t x^t + 2c_{t-2}x^{t-2} + \cdots + 2c_s x^s.$$

Let

$$F(x) = a_k x^k + a_{k-1}x^{k-1} + \cdots + a_2 x^2 + a_1 x + a_0.$$

Since $s < t \leq n/3 < m$ we have that $s + 4 \leq m$.

Investigate the coefficients of x^s and x^{s+2} in

$$vP_{n,m}(x) = F(G(x)) = a_k G(x)^k + a_{k-1}G(x)^{k-1} + \cdots + a_1 G(x) + a_0. \tag{6}$$

Since $s + 2 < m$ in the polynomials $vP_{n,m}(x) = E_n(x) + cE_m(x)$ these coefficients are 0. On the other hand, one can observe that x^s occurs only in the term $a_1 G(x)$ and so $a_1 c_s = 0$. This means that $a_1 = 0$ and so

$$vP_{n,m}(x) = F(G(x)) = a_k G(x)^k + \cdots + a_3 G(x)^3 + a_2 G(x)^2 + a_0. \tag{7}$$

Since x^{s+2} appears only in the term $a_2 G(x)^2$ thus $2a_2 c_2 c_s = 0$.

If $a_2 = 0$ we obtain from (7) that the coefficients of x^5, x^4, x^3, x^2 and x in $F(G(x))$ are zero. This yields that $P_{n,m}^{(i)}(0) = 0$ for $i = 1, \ldots, 5$. Further, by Lemma 1

$$P_{n,m}^{(j)}(0) = P_{n,m}^{(j)}(1) = 0, \text{ if } j \text{ is odd and } j \neq m, n;$$

$$P_{n,m}^{(j)}\left(\frac{1}{2}\right) = 0, \text{ if } j \text{ is even.}$$

Applying the above, we can study the number of zeros of the polynomials $P_{n,m}^{(j)}(x)$ in the interval $[0,1]$ for $j = 1, 2, \ldots, m + 1$. In the following table we use only the Rolle's theorem.

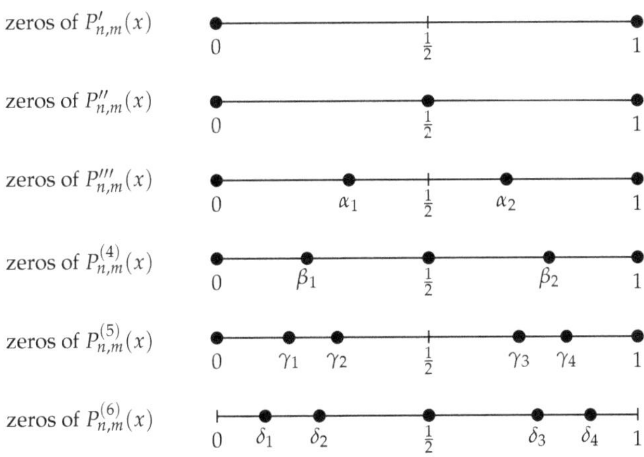

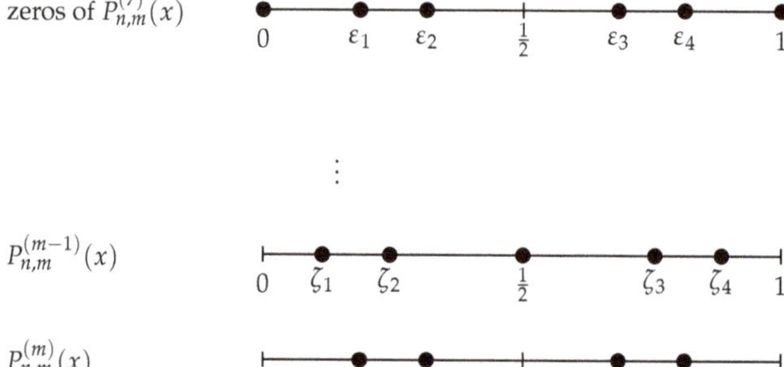

But

$$P_{n,m}^{(m+1)}(x) = \frac{n!}{(n-m-1)!} E_{n-m-1}(x)$$

whose the only zero in the interval $[0,1]$ is $1/2$. This contradiction gives that $a_2 \neq 0$.

If $c_2 = 0$ then from $G(x) = c_t x^t + \cdots + c_3 x^3$ and (7) one can deduce that

$$P_{n,m}^{(j)}(0) \text{ for } j = 1,2,3,4,5.$$

The above argument that we used in the case $a_2 = 0$ shows that this impossible.

Finally, consider the case when $\mathbf{s} = \mathbf{1}$. Let $c = A/B$, where A and $B \neq 0$ are relatively prime integers. From (4) we know that

$$g(x) - g(-x) = 2c_t x^t + 2c_{t-2} x^{t-2} + \cdots + 2c_3 x^3 + 2c_1 x = dxh(x), \tag{8}$$

where the polynomial $h(x)$ is even and divides the polynomial $Bx^{n-m} + A$ in $\mathbb{Q}[x]$. If we write $h(x)$ as a product of a rational number a/b and a primitive polynomial $H(x) = h_r x^r + h_{r-2} x^{r-2} + \cdots + h_2 x^2 + h_0$ we have that

$$Bx^{n-m} + A = H(x)u(x), \tag{9}$$

where $u(x) = u_q x^q + u_{q-2} x^{q-2} + \cdots + u_2 x^2 + u_0$ is a primitive polynomial. We obtain from (8) and (9) that

$$(2c_t x^t + 2c_{t-2} x^{t-2} + \cdots + 2c_3 x^3 + 2c_1 x)u(x) = \frac{a}{b} d(Bx^{n-m+1} + Ax). \tag{10}$$

Let $c_t = wc'_t, c_{t-2} = wc'_{t-2}, \ldots, c_3 = wc'_3$ and $c_1 = wc'_1$, where w denotes the greatest common divisor of the integers $c_t, c_{t-2}, \ldots, c_3, c_1$. Then

$$2w(c'_t x^t + c'_{t-2} x^{t-2} + \cdots + c'_3 x^3 + c'_1 x)u(x) = \frac{a}{b} d(Bx^{n-m+1} + Ax), \tag{11}$$

which yields that $2w = (a/b)d$ and

$$(c'_t x^t + c'_{t-2} x^{t-2} + \cdots + c'_1 x)(u_q x^q + u_{q-2} x^{q-2} + \cdots + u_0) = Bx^{n-m+1} + Ax. \tag{12}$$

It follows from Lemma 6 that if p is an odd prime which divides w then p divides $c_t, c_{t-1}, \ldots, c_2, c_1, c_0$ which is not possible since $g(x)$ is a primitive polynomial. Thus $w = 2^a$ for some non-negative integer a. Now assume that p is a prime which divides c'_t and $j \geq 1$ is the greatest odd index for which

$$p|c'_t, c'_{t-2}, \ldots, c'_{j+2} \quad \text{and} \quad p \nmid c'_j. \tag{13}$$

On the right hand side of (12) the coefficient of x^α equals 0 apart from when $\alpha = q + t = n - m + 1$ or $\alpha = 1$. Thus

$$c'_j u_q + c'_{j+2} u_{q-2} + c'_{j+4} u_{q-4} + \cdots = 0$$

which means that $p|u_q$.

Similarly,

$$c'_{j-2} u_q + c'_j u_{q-2} + c'_{j+2} u_{q-4} + \cdots = 0$$

from which we get that $p|u_{q-2}$. Continuing the process one can deduce that

$$p|u_q, u_{q-2}, \ldots, u_2.$$

Further, if $j > 1$ then

$$c'_j u_0 + c'_{j-2} u_2 + \ldots = 0$$

and so $p|u_0$ contradicting that the polynomial $u(x)$ is a primitive polynomial. It follows from the above that j must be 1 and so

$$p|c'_t, c'_{t-2}, \ldots, c'_3, u_q, u_{q-2}, \ldots, u_2 \quad \text{and} \quad p \nmid c'_1, u_0. \tag{14}$$

If p is an odd prime then from the above and Lemma 6 we have that

$$p|c_t, c_{t-1}, c_{t-2} \ldots, c_3, c_2 \text{ and } p \nmid c_1, c_0. \tag{15}$$

Now let $U(x) = c_t x^t + c_{t-1} x^{t-1} + \cdots + c_2 x^2$ and $V(x) = c_1 x + c_0$. Then $g(x) = U(x) + V(x)$ and for $j = 0, 1, \ldots, k$

$$g(x)^j = \sum_{i=0}^{j} \binom{j}{i} U(x)^{j-i} V(x)^i \equiv V(x)^j \mod (p). \tag{16}$$

We know that $m > n/3 \geq k$ and so the coefficients of $x^k, x^{k-2}, \ldots, x^3, x$ are zeros in $P_{n,m}(x)$ and so in

$$f(g(x)) = b_k g(x)^k + b_{k-2} g(x)^{k-2} + \cdots + b_3 g(x)^3 + b_1 g(x), \text{ too.} \tag{17}$$

Now one can infer from (16) and (17) that $0 \equiv b_k c_1^k \mod (p)$ which yields $p|b_k$. Comparing coefficient of x^{k-2} we have that $0 \equiv b_{k-2} c_1^{k-2} \mod (p)$ from which we obtain $p|b_{k-2}$. Continuing the process it is easy to see that $p|b_{k-4}, \ldots, b_3, b_1$ which contradicts the fact that $f(x)$ is a primitive polynomial. This means that c'_t and c_t must be powers of two.

Now suppose that p is a prime with $p|u_q$ and $p \nmid c'_t$. Using again that on the right hand side of (12) the coefficient of x^α equals 0 apart from when $\alpha = n - m + 1$ or 1. From $u_q c'_{t-2} + u_{q-2} c'_t = 0$ we obtain p divides u_{q-2}. From $u_q c'_{t-4} + u_{q-2} c'_{t-2} + u_{q-4} c'_t = 0$ we obtain p divides u_{q-4}. It follows similarly that $p|u_{q-6}, \ldots, u_2$. Finally, from $c'_t u_0 + c'_{t-2} u_2 + \cdots = 0$ we get that p divides u_0 which contradicts that the polynomial $u(x)$ is a primitive polynomial. This means that u_q must be a power of two. Since $B = c'_t u_q$ this contradicts to our assumption that B is not a power of two.

4. Concluding Remarks

It is a very hard problem to characterize the general decomposition of an infinite sequence of polynomials $f_n(x)$. The first theorem was proved for Bernoulli polynomials. For other results see our

Introduction. A harder question is to describe the decomposition of the sum of two polynomials. There are only a few results in this direction, mainly for the rational linear combination of two Bernoulli and Euler polynomials in the form $B_n(x) + cB_{n-2}$ and $E_n(x) + cE_{n-2}(x)$, respectively. This paper contains the first theorem concerning the decomposition of the linear combination of two Euler polynomials $E_m x + cE_n(x)$ with "almost" independent parameters m and n.

Author Contributions: All the authors contributed equally to the conception of the idea, implementing and analyzing the experimental results, and writing the manuscript.

Funding: This research Supported in part by the Hungarian Academy of Siences, and NKFIH/OTKA grant K128088.

Conflicts of Interest: The authors declare no conflict of interest.

References

1. Abramowitz, M.; Stegun, I. *Handbook of Mathematical Functions with Formulas, Graphs and Mathematical Tables*; Dover: New York, NY, USA, 1972.
2. Brillhart, J. On the Euler and Bernoulli polynomials. *J. Reine Angew. Math.* **1969**, *234*, 45–64.
3. Koblitz, N. *p-adic Numbers, p-adic Analysis, and Zeta-Functions*, 2nd ed.; Graduate Texts in Mathematics; Springer: New York, NY, USA, 1984.
4. Bagarello, F.; Trapani, C.; Triolo, S. Representable states on quasilocal quasi*-algebras. *J. Math. Phys.* **2011**, *52*, 013510. [CrossRef]
5. Ballentine, L.E.; McRae, S.M. Moment equations for probability distributions in classical and quantum mechanics. *Phys. Rev. A* **1998**, *58*, 1799–1809. [CrossRef]
6. Trapani, C.; Triolo, S. Representations of modules over a*-algebra and related seminorms. *Stud. Math.* **2008**, *184*, 133–148. [CrossRef]
7. Triolo, S. WQ* algebras of measurable operators. *Indian J. Pure Appl. Math.* **2012**, *43*, 601–617. [CrossRef]
8. Dujella, A.; Gusíc, I. Indecomposability of polynomials and related Diophantine equations. *Q. J. Math.* **2006**, *57*, 193–201. [CrossRef]
9. Dujella, A.; Gusíc, I. Decomposition of a recursive family of polynomials. *Monatshefte für Mathematik* **2007**, *152*, 97–104. [CrossRef]
10. Dujella, A.; Gusic, I.; Tichy, R.F. On the indecomposability of polynomials. *Osterreich. Akad. Wiss. Math.-Natur. Kl. Sitzungsber. II* **2005**, *214*, 81–88. [CrossRef]
11. Dujella, A.; Tichy, R.F. Diophantine equations for second-order recursive sequences of polynomials. *Quart. J. Math.* **2001**, *52*, 161–169. [CrossRef]
12. Kreso, D.; Tichy, R.F. Functional composition of polynomials: Indecomposability, Diophantine equations and lacunary polynomials. *Graz. Math. Ber.* **2015**, *363*, 143–170.
13. Bilu, Y.; Brindza, B.; Kirschenhofer, P.; Pintér, Á.; Tichy, R. Diophantine Equations and Bernoulli Polynomials. *Compositio Math.* **2002**, *131*, 173–188. [CrossRef]
14. Kreso, D.; Rakaczki, C. Diophantine equations with Euler polynomials. *Acta Arith.* **2013**, *161*, 267–281. [CrossRef]
15. Pintér, Á.; Rakaczki, C. On the decomposability of linear combinations of Bernoulli. *Monatshefte Math.* **2016**, *180*, 631–648. [CrossRef]
16. Pintér, Á.; Rakaczki, C. On the decomposability of the linear combinations of Euler polynomials. *Miskolc Math. Notes* **2017**, *18*, 407–415. [CrossRef]

© 2019 by the authors. Licensee MDPI, Basel, Switzerland. This article is an open access article distributed under the terms and conditions of the Creative Commons Attribution (CC BY) license (http://creativecommons.org/licenses/by/4.0/).

Article

Structure of Approximate Roots Based on Symmetric Properties of (p,q)-Cosine and (p,q)-Sine Bernoulli Polynomials

Cheon Seoung Ryoo [1] and Jung Yoog Kang [2],*

[1] Department of Mathematics, Hanman University, Daejeon 10216, Korea; ryoocs@hnu.kr
[2] Department of Mathematics Education, Silla University, Busan 469470, Korea
* Correspondence: jykang@silla.ac.kr

Received: 13 May 2020; Accepted: 27 May 2020; Published: 30 May 2020

Abstract: This paper constructs and introduces (p,q)-cosine and (p,q)-sine Bernoulli polynomials using (p,q)-analogues of $(x+a)^n$. Based on these polynomials, we discover basic properties and identities. Moreover, we determine special properties using (p,q)-trigonometric functions and verify various symmetric properties. Finally, we check the symmetric structure of the approximate roots based on symmetric polynomials.

Keywords: (p,q)-cosine Bernoulli polynomials; (p,q)-sine Bernoulli polynomials; (p,q)-numbers; (p,q)-trigonometric functions

MSC: 11B68; 11B75; 33A10

---

1. Introduction

In 1991, (p,q)-calculus including (p,q)-number with two independent variables p and q, was first independently considered [1,2]. Throughout this paper, the sets of natural numbers, integers, real numbers and complex numbers are denoted by $\mathbb{N}, \mathbb{Z}, \mathbb{R}$ and $\mathbb{C}$, respectively.

For any $n \in \mathbb{N}$, the (p,q)-number is defined by the following:

$$[n]_{p,q} = \frac{p^n - q^n}{p - q}, \qquad \text{where} \quad |p/q| < 1, \tag{1}$$

which is a natural generalization of the q-number. From Equation (1), we note that $[n]_{p,q} = [n]_{q,p}$.

Many physical and mathematical problems have led to the necessity of studying (p,q)-calculus. Since 1991, many mathematicians and physicists have developed (p,q)-calculus in several different research areas. For example, in 1994, [3] introduced (p,q)-hypergeometric functions. Three years later, [3,4] derived related preliminary results by considering a more general (p,q)-hypergeometric series and Burban's (p,q)-hypergeometric series, respectively. In 2005, based on the (p,q)-numbers, [5] studied about (p,q)-hypergeometric series and discovered results corresponding to the (p,q)-extensions of known q-identities. Moreover, [6] established properties similar to the ordinary and q-binomial coefficients after developing the (p,q)-hypergeometric series in 2008. About seven years later, [7] introduced (p,q)-gamma and (p,q)-beta functions, which are generalizations of the gamma and beta functions.

The different variations of Bernoulli polynomials, q-Bernoulli polynomials and (p,q)-Bernoulli polynomials are illustrated in the diagram below. Kim, Ryoo and many mathematicians have studied the first and second rows of the polynomials in the diagram(see [8–12]). These studies began producing valuable results in areas related to number theory and combinatorics.

The main idea is to use property of (p,q)-numbers and combine (p,q)-trigonometric functions. From this idea, we construct (p,q)-cosine and (p,q)-sine Bernoulli polynomials. Investigating the various explicit identities for (p,q)-cosine and (p,q)-sine Bernoulli polynomials in the diagram's third row is the main goal of this paper.

$$\frac{t}{e^t-1}e^{tx} = \sum_{n=0}^{\infty} B_n(x)\frac{t^n}{n!}$$
(Bernoulli polynomials)

$$\frac{t}{e^t-1}e^{tx}\cos(ty) = \sum_{n=0}^{\infty} B_n^{(C)}(x,y)\frac{t^n}{n!}$$
(cosine Bernoulli polynomials)

$$\frac{t}{e^t-1}e^{tx}\sin(ty) = \sum_{n=0}^{\infty} B_n^{(S)}(x,y)\frac{t^n}{n!}$$
(sine Bernoulli polynomials)

$$\frac{t}{e_q(t)-1}e_q(tx) = \sum_{n=0}^{\infty} B_{n,q}(x)\frac{t^n}{n!}$$
(q-Bernoulli polynomials)

$$\frac{t}{e_q(t)-1}e_q(tx)COS_q(ty) = \sum_{n=0}^{\infty} {}_CB_{n,q}(x,y)\frac{t^n}{n!}$$
(q-cosine Bernoulli polynomials)

$$\frac{t}{e_q(t)-1}e_q(tx)SIN_q(ty) = \sum_{n=0}^{\infty} {}_SB_{n,q}(x,y)\frac{t^n}{n!}$$
(q-sine Bernoulli polynomials)

$$\frac{t}{e_{p,q}(t)-1}e_{p,q}(tx) = \sum_{n=0}^{\infty} B_{n,p,q}(x)\frac{t^n}{n!}$$
((p,q)-Bernoulli polynomials)

$$\frac{t}{e_{p,q}(t)-1}e_{p,q}(tx)COS_{p,q}(ty) = \sum_{n=0}^{\infty} {}_CB_{n,p,q}(x,y)\frac{t^n}{n!}$$
((p,q)-cosine Bernoulli polynomials)

$$\frac{t}{e_{p,q}(t)-1}e_{p,q}(tx)SIN_{p,q}(ty) = \sum_{n=0}^{\infty} {}_SB_{n,p,q}(x,y)\frac{t^n}{n!}$$
((p,q)-sine Bernoulli polynomials)

Due to their importance, the classical Bernoulli, Euler, and Genocchi polynomials have been studied extensively and are well-known. Mathematicians have studied these polynomials through various mathematical applications including finite difference calculus, p-adic analytic number theory, combinatorial analysis and number theory. Many mathematicians are familiar with the theorems and definitions of classical Bernoulli, Euler, and Genocchi polynomials. Based on the theorems and definitions, it is significant to study these properties in various ways by the combining with Bernoulli, Euler, and Genocchi polynomials. Mathematicians are studying the extended versions of these polynomials and are researching new polynomials by combining mathematics with other fields, such as physics or engineering (see [9–14]). The definition of Bernoulli polynomials combined with (p,q)-numbers follows:

Definition 1. *The (p,q)-Bernoulli numbers, $B_{n,p,q}$, and polynomials, $B_{n,p,q}(z)$, can be expressed as follows (see [8])*

$$\sum_{n=0}^{\infty} B_{n,p,q}\frac{t^n}{[n]_{p,q}!} = \frac{t}{e_{p,q}(t)-1}, \quad \sum_{n=0}^{\infty} B_{n,p,q}(z)\frac{t^n}{[n]_{p,q}!} = \frac{t}{e_{p,q}(t)-1}e_{p,q}(tz). \tag{2}$$

In [11], we confirmed the properties of q-cosine and q-sine Bernoulli polynomials. Their definitions and representative properties are as follows.

Definition 2. *The q-cosine Bernoulli polynomials ${}_CB_{n,q}(x,y)$ and q-sine Bernoulli polynomials ${}_SB_{n,q}(x,y)$ are defined by the following:*

$$\sum_{n=0}^{\infty} {}_CB_{n,q}(x,y)\frac{t^n}{n!} = \frac{t}{e_q(t)-1}e_q(tx)COS_q(ty), \quad \sum_{n=0}^{\infty} {}_SB_n(x,y)\frac{t^n}{n!} = \frac{t}{e_q(t)-1}e_q(tx)SIN_q(ty). \tag{3}$$

Theorem 1. *For $x, y \in \mathbb{R}$, we have the following:*

$$(i) \quad \begin{cases} {}_cB_{n,q}((x \oplus r)_q, y) + {}_sB_{n,q}((x \ominus r)_q, y) \\[2mm] = \sum_{k=0}^{n} \begin{bmatrix} n \\ k \end{bmatrix}_q q^{\binom{n-k}{2}} r^{n-k} \left({}_cB_{k,q}(x, y) + (-1)^{n-k} {}_sB_{k,q}(x, y) \right) \\[3mm] {}_sB_{n,q}((x \oplus r)_q, y) + {}_cB_{n,q}((x \ominus r)_q, y) \\[2mm] = \sum_{k=0}^{n} \begin{bmatrix} n \\ k \end{bmatrix}_q q^{\binom{n-k}{2}} r^{n-k} \left({}_sB_{k,q}(x, y) + (-1)^{n-k} {}_cB_{k,q}(x, y) \right) \end{cases}$$

$$(ii) \quad \begin{cases} \frac{\partial}{\partial x} {}_cB_{n,q}(x, y) = [n]_q {}_cB_{n-1,q}(x, y), \quad \frac{\partial}{\partial y} {}_cB_{n,q}(x, y) = -[n]_q {}_sB_{n-1,q}(x, qy) \\[2mm] \frac{\partial}{\partial x} {}_sB_{n,q}(x, y) = [n]_q {}_sB_{n-1,q}(x, y), \quad \frac{\partial}{\partial y} {}_sB_{n,q}(x, y) = [n]_q {}_cB_{n-1,q}(x, qy) \end{cases}$$

$$(4)$$

The main goal of this paper is to identify the properties of (p, q)-cosine and (p, q)-sine Bernoulli polynomials. In Section 2, we review some definitions and theorem of (p, q)-calculus. In Section 3, we introduce (p, q)-cosine and (p, q)-sine Bernoulli polynomials. Using the properties of exponential functions and trigonometric functions associated with (p, q)-numbers, we determine the various properties and identities of the polynomials. Section 4 presents the investigation of the symmetric properties of (p, q)-cosine and (p, q)-sine Bernoulli polynomials in different forms and based on the symmetric polynomials, we check the symmetric structure of the approximate roots.

2. Preliminaries

In this section, we introduce definitions and preliminary facts that are used throughout this paper (see [6,12–20]).

Definition 3. *For $n \geq k$, the Gaussian binomial coefficients are defined by the following:*

$$\begin{bmatrix} m \\ r \end{bmatrix}_{p,q} = \frac{[n]_{p,q}!}{[n-k]_{p,q}! [k]_{p,q}!}, \tag{5}$$

where m and r are non-negative integers.

We note that $[n]_{p,q}! = [n]_{p,q}[n-1]_{p,q} \cdots [2]_{p,q}[1]_{p,q}$, where $n \in \mathbb{N}$. For $r = 0$, the value of the equation is 1, because both the numerator and denominator are empty products. Moreover, (p, q)-analogues of the binomial formula exist, and this definition has numerous properties.

Definition 4. *The (p, q)-analogues of $(x - a)^n$ and $(x + a)^n$ are defined by the following:*

$$(i) \quad (x \ominus a)_{p,q}^{n} = \begin{cases} 1, & \text{if } n = 0 \\ (x - a)(px - qa) \cdots (p^{n-1}x - q^{n-1}a), & \text{if } n \geq 1 \end{cases}$$

$$(ii) \quad (x \oplus a)_{p,q}^{n} = \begin{cases} 1, & \text{if } n = 0 \\ (x + a)(px + qa) \cdots (p^{n-2}x + q^{n-2}a)(p^{n-1}x + q^{n-1}a), & \text{if } n \geq 1 \end{cases}$$

$$= \sum_{k=0}^{n} \begin{bmatrix} n \\ k \end{bmatrix}_{p,q} p^{\binom{k}{2}} q^{\binom{n-k}{2}} x^k a^{n-k}. \tag{6}$$

Definition 5. *We express the two forms of (p, q)-exponential functions as follows:*

$$e_{p,q}(x) = \sum_{n=0}^{\infty} p^{\binom{n}{2}} \frac{x^n}{[n]_{p,q}!}, \qquad E_{p,q}(x) = \sum_{n=0}^{\infty} q^{\binom{n}{2}} \frac{t^n}{[n]_{p,q}!}. \tag{7}$$

From Equation (7), we determine an important property, $e_{p,q}(x)E_{p,q}(-x) = 1$. Moreover, Duran, Acikgos, and Araci defined $\widetilde{e}_{p,q}(x)$ in [17] as follows:

$$\widetilde{e}_{p,q}(x) = \sum_{n=0}^{\infty} \frac{x^n}{[n]_{p,q}!}. \tag{8}$$

From Equations (8) and (6), we remark $e_{p,q}(x)E_{p,q}(y) = \widetilde{e}_{p,q}(x \oplus y)_{p,q}$.

Definition 6. *For $x \neq 0$, the (p,q)-derivative of a function f with respect to x is defined by the following:*

$$D_{p,q}f(x) = \frac{f(px) - f(qx)}{(p-q)x}, \tag{9}$$

where $(D_{p,q}f)(0) = f'(0)$, which prove that f is differentiable at 0. Moreover, it is evident that $D_{p,q}x^n = [n]_{p,q}x^{n-1}$.

Definition 7. *Let $i = \sqrt{-1} \in \mathbb{C}$. Then the (p,q)-trigonometric functions are defined by the following:*

$$
\begin{aligned}
sin_{p,q}(x) &= \frac{e_{p,q}(ix) - e_{p,q}(-ix)}{2i}, & SIN_{p,q}(x) &= \frac{E_{p,q}(ix) - E_{p,q}(-ix)}{2i} \\
cos_{p,q}(x) &= \frac{e_{p,q}(ix) + e_{p,q}(-ix)}{2}, & COS_{p,q}(x) &= \frac{E_{p,q}(ix) + E_{p,q}(-ix)}{2},
\end{aligned} \tag{10}
$$

where, $SIN_{p,q}(x) = sin_{p^{-1},q^{-1}}(x)$ and $COS_{p,q}(x) = cos_{p^{-1},q^{-1}}(x)$.

In the same way as well-known Euler expressions using exponential functions, we define the (p,q)-analogues of hyperbolic functions and find several formulae (see [3,5,17]).

Theorem 2. *The following relationships hold:*

$$
\begin{aligned}
&(i) \quad sin_{p,q}(x)COS_{p,q}(x) = cos_{p,q}(x)SIN_{p,q}(x) \\
&(ii) \quad e_{p,q}(x) = cosh_{p,q}(x)sinh_{p,q}(x) \\
&(iii) \quad E_{p,q}(x) = COSH_{p,q}(x)SINH_{p,q}(x).
\end{aligned} \tag{11}
$$

From Definition 7 and Theorem 2, we note that $cosh_{p,q}(x)COSH_{p,q}(x) - sinh_{p,q}(x)SINH_{p,q}(x) = 1$.

3. Several Basic Properties of (p,q)-Cosine and (p,q)-Sine Bernoulli Polynomials

We look for Lemma 1 and Theorem 3 in order to introduce (p,q)-cosine and (p,q)-sine Bernoulli polynomials. From the definitions of the (p,q)-cosine and (p,q)-sine Bernoulli polynomials, we search for a variety of properties. We also find relationships with other polynomials using properties of (p,q)-trigonometric functions or other methods.

Lemma 1. *For $y \in \mathbb{R}$ and $i = \sqrt{-1}$, we have the following:*

$$
\begin{aligned}
&(i) \quad E_{p,q}(ity) = COS_{p,q}(ty) + iSIN_{p,q}(ty), \\
&(ii) \quad E_{p,q}(-ity) = COS_{p,q}(ty) - iSIN_{p,q}(ty).
\end{aligned} \tag{12}
$$

Proof.

(i) $E_{p,q}(ity)$ can be expressed using the (p,q)-cosine and (p,q)-sine functions as

$$E_{p,q}(ity) = \frac{E_{p,q}(ity) + E_{p,q}(-ity)}{2} + \frac{E_{p,q}(ity) - E_{p,q}(-ity)}{2} \tag{13}$$
$$= COS_{p,q}(ty) + iSIN_{p,q}(ty).$$

(ii)   By substituting $-ity$ instead of (i), we obtain the following:

$$E_{p,q}(-ity) = \frac{E_{p,q}(-ity) + E_{p,q}(-ity)}{2} - \frac{E_{p,q}(-ity) - E_{p,q}(-ity)}{2} \tag{14}$$
$$= COS_{p,q}(ty) - iSIN_{p,q}(ty).$$

Therefore, we complete the proof of Lemma 1. $\square$

We note the following relations between $e_{p,q}$, $E_{p,q}$ and $\tilde{e}_{p,q}$.

$$(i) \quad e_{p,q}(x)E_{p,q}(y) = \sum_{n=0}^{\infty} \frac{(x \oplus y)_{p,q}^{n}}{[n]_{p,q}!} = \tilde{e}_{p,q}(x \oplus y)_{p,q},$$

$$(ii) \quad e_{p,q}(x)E_{p,q}(-y) = \sum_{n=0}^{\infty} \frac{(x \ominus y)_{p,q}^{n}}{[n]_{p,q}!} = \tilde{e}_{p,q}(x \ominus y)_{p,q}. \tag{15}$$

Theorem 3. *Let* $x, y \in \mathbb{R}$, $i = \sqrt{-1}$, *and* $|q/p| < 1$. *Then, we have*

$$(i) \quad \sum_{n=0}^{\infty} \sum_{k=0}^{n} \begin{bmatrix} n \\ k \end{bmatrix}_{p,q} \left(\frac{(x \oplus iy)_{p,q}^{k} + (x \ominus iy)_{p,q}^{k}}{2} \right) B_{n-k,p,q} \frac{t^n}{[n]_{p,q}!}$$

$$= \frac{t}{e_{p,q}(t) - 1} e_{p,q}(tx) COS_{p,q}(ty),$$

$$(ii) \quad \sum_{n=0}^{\infty} \sum_{k=0}^{n} \begin{bmatrix} n \\ k \end{bmatrix}_{p,q} \left(\frac{(x \oplus iy)_{p,q}^{k} - (x \ominus iy)_{p,q}^{k}}{2i} \right) B_{n-k,p,q} \frac{t^n}{[n]_{p,q}!} \tag{16}$$

$$= \frac{t}{e_{p,q}(t) - 1} e_{p,q}(tx) SIN_{p,q}(ty).$$

Proof.

(i)   We note that

$$\sum_{n=0}^{\infty} B_{n,p,q} \frac{t^n}{[n]_{p,q}!} - \frac{t}{e_{p,q}(t) - 1}. \tag{17}$$

We find the following by multiplying $\tilde{e}_{p,q}\left(t(x \oplus y)_{p,q}\right)$ in both sides of Equation (17).

$$\sum_{n=0}^{\infty} B_{n,p,q} \frac{t^n}{[n]_{p,q}!} \tilde{e}_{p,q}\left(t(x \oplus y)_{p,q}\right) = \frac{t}{e_{p,q}(t) - 1} \tilde{e}_{p,q}\left(t(x \oplus y)_{p,q}\right). \tag{18}$$

The left-hand side of Equation (18) can be changed into

$$\sum_{n=0}^{\infty} B_{n,p,q} \frac{t^n}{[n]_{p,q}!} \tilde{e}_{p,q}\left(t(x \oplus y)_{p,q}\right) = \sum_{n=0}^{\infty} B_{n,p,q} \frac{t^n}{[n]_{p,q}!} \sum_{n=0}^{\infty} (x \oplus y)_{p,q}^{n} \frac{t^n}{[n]_{p,q}!}$$

$$= \sum_{n=0}^{\infty} \left(\sum_{k=0}^{n} \begin{bmatrix} n \\ k \end{bmatrix}_{p,q} (x \oplus y)_{p,q}^{k} B_{n-k,p,q} \right) \frac{t^n}{[n]_{p,q}!}, \tag{19}$$

and by using Lemma 1 (i) on the right-hand side of Equation (18), we yield

$$\frac{t}{e_{p,q}(t)-1}\tilde{e}_{p,q}\left(t(x\oplus y)_{p,q}\right) = \frac{t}{e_{p,q}(t)-1}e_{p,q}(x)E_{p,q}(y)$$

$$= \frac{te_{p,q}(x)}{e_{p,q}(t)-1}\left(COS_{p,q}(ty) + iSIN_{p,q}(ty)\right). \tag{20}$$

From Equations (19) and (20), we derive the following:

$$\sum_{n=0}^{\infty}\left(\sum_{k=0}^{n}\begin{bmatrix}n\\k\end{bmatrix}_{p,q}(x\oplus y)_{p,q}^{k}B_{n-k,p,q}\right)\frac{t^n}{[n]_{p,q}!} = \frac{te_{p,q}(x)}{e_{p,q}(t)-1}\left(COS_{p,q}(ty) + iSIN_{p,q}(ty)\right). \tag{21}$$

We obtain the equation below for (p,q)-Bernoulli numbers using a similar method.

$$\sum_{n=0}^{\infty}\left(\sum_{k=0}^{n}\begin{bmatrix}n\\k\end{bmatrix}_{p,q}(x\ominus iy)_{p,q}^{k}B_{n-k,p,q}\right)\frac{t^n}{[n]_{p,q}!} = \frac{te_{p,q}(tx)}{e_{p,q}(t)-1}\left(COS_{p,q}(ty) - iSIN_{p,q}(ty)\right). \tag{22}$$

By using Equations (21) and (22), we have

$$\sum_{n=0}^{\infty}\sum_{k=0}^{n}\begin{bmatrix}n\\k\end{bmatrix}_{p,q}\left(\frac{(x\oplus iy)_{p,q}^{k}+(x\ominus iy)_{p,q}^{k}}{2}\right)B_{n-k,p,q}\frac{t^n}{[n]_{p,q}!} = \frac{t}{e_{p,q}(t)-1}e_{p,q}(tx)COS_{p,q}(ty) \tag{23}$$

and

$$\sum_{n=0}^{\infty}\sum_{k=0}^{n}\begin{bmatrix}n\\k\end{bmatrix}_{p,q}\left(\frac{(x\oplus iy)_{p,q}^{k}-(x\ominus iy)_{p,q}^{k}}{2i}\right)B_{n-k,p,q}\frac{t^n}{[n]_{p,q}!} = \frac{t}{e_{p,q}(t)-1}e_{p,q}(tx)SIN_{p,q}(ty). \tag{24}$$

Therefore, we can conclude the required results. □

Thus, we are ready to introduce (p,q)-cosine and (p,q)-sine Bernoulli polynomials using Lemma 1 and Theorem 3.

Definition 8. *Let $|p/q| < 1$ and $x,y \in \mathbb{R}$. Then (p,q)-cosine and (p,q)-sine Bernoulli polynomials are respectively defined by the following:*

$$\sum_{n=0}^{\infty}{}_{C}B_{n,p,q}(x,y)\frac{t^n}{[n]_{p,q}!} = \frac{t}{e_{p,q}(t)-1}e_{p,q}(tx)COS_{p,q}(ty),$$

and

$$\sum_{n=0}^{\infty}{}_{S}B_{n,p,q}(x,y)\frac{t^n}{[n]_{p,q}!} = \frac{t}{e_{p,q}(t)-1}e_{p,q}(tx)SIN_{p,q}(ty). \tag{25}$$

From Definition 8, we determine q-cosine and q-sine Bernoulli polynomials when $|q| < 1$ and $p = 1$. In addition, we observe cosine Bernoulli polynomials and sine Bernoulli polynomials for $q \to 1$ and $p = 1$.

Corollary 1. *From Theorem 3 and Definition 8, the following holds*

$$
\begin{aligned}
(i) \quad {}_C B_{n,p,q}(x,y) &= \sum_{k=0}^{n} \begin{bmatrix} n \\ k \end{bmatrix}_{p,q} \left(\frac{(x \oplus iy)_{p,q}^{k} + (x \ominus iy)_{p,q}^{k}}{2} \right) B_{n-k,p,q}, \\
(ii)\,{}_S B_{n,p,q}(x,y) &= \sum_{k=0}^{n} \begin{bmatrix} n \\ k \end{bmatrix}_{p,q} \left(\frac{(x \oplus iy)_{p,q}^{k} - (x \ominus iy)_{p,q}^{k}}{2i} \right) B_{n-k,p,q},
\end{aligned}
\tag{26}
$$

where $B_{n,p,q}$ denotes the (p,q)-Bernoulli numbers.

Example 1. *From Definition 8, a few examples of ${}_C B_{n,p,q}(x,y)$ and ${}_S B_{n,p,q}(x,y)$ are the follows:*

$$
\begin{aligned}
{}_C B_{0,p,q}(x,y) &= 0 \\
{}_C B_{1,p,q}(x,y) &= px \\
{}_C B_{2,p,q}(x,y) &= p^2 x^2 - qy^2 \\
{}_C B_{3,p,q}(x,y) &= p^3 x^3 - pq(p^2 + pq + q^2)xy^2 \\
{}_C B_{4,p,q}(x,y) &= p^4 x^4 - p^2 q(p^2 + q^2)(p^2 + pq + q^2)x^2 y^2 + q^6 y^4, \\
&\cdots .
\end{aligned}
\tag{27}
$$

and

$$
\begin{aligned}
{}_S B_{0,p,q}(x,y) &= 0 \\
{}_S B_{1,p,q}(x,y) &= \frac{y}{p+q} \\
{}_S B_{2,p,q}(x,y) &= \frac{pxy}{p^2 + pq + q^2} \\
{}_S B_{3,p,q}(x,y) &= \frac{y \left(\frac{p^2 x^2}{p^2 + q^2} - q^3 y^2 \right)}{p+q} \\
{}_S B_{4,p,q}(x,y) &= p(p-q)xy \left(\frac{p^2 x^2}{p^5 - q^5} + \frac{q^3(p^2 + q^2)y^2}{-p^3 + q^3} \right), \\
&\cdots .
\end{aligned}
\tag{28}
$$

Definition 9. *Let $|p/q| < 1$. Then, we define*

$$
\sum_{n=0}^{\infty} C_{n,p,q}(x,y) \frac{t^n}{[n]_{p,q}!} = e_{p,q}(tx) COS_{p,q}(ty), \quad \sum_{n=0}^{\infty} S_{n,p,q}(x,y) \frac{t^n}{[n]_{p,q}!} = e_{p,q}(tx) SIN_{p,q}(ty).
\tag{29}
$$

Theorem 4. *Let k be a nonnegative integer and $|p/q| < 1$. Then, we have*

$$
\begin{aligned}
(i) \quad {}_C B_{n,p,q}(x,y) &= \sum_{k=0}^{n} \begin{bmatrix} n \\ k \end{bmatrix}_{p,q} B_{n-k,p,q} C_{k,p,q}(x,y), \\
(ii) \quad {}_S B_{n,p,q}(x,y) &= \sum_{k=0}^{n} \begin{bmatrix} n \\ k \end{bmatrix}_{p,q} B_{n-k,p,q} S_{k,p,q}(x,y),
\end{aligned}
\tag{30}
$$

where $B_{n,p,q}$ is the (p,q)-Bernoulli numbers.

Proof.

(i) Using the generating function of the (p,q)-cosine Bernoulli polynomials and Definition 9, we find

$$\sum_{n=0}^{\infty} {}_{C}B_{n,p,q}(x,y)\frac{t^n}{[n]_{p,q}!} = \sum_{n=0}^{\infty} B_{n,p,q}\frac{t^n}{[n]_{p,q}!}\sum_{n=0}^{\infty} C_{n,p,q}(x,y)\frac{t^n}{[n]_{p,q}!}$$

$$= \sum_{n=0}^{\infty}\left(\sum_{k=0}^{n}\begin{bmatrix}n\\k\end{bmatrix}_{p,q} B_{n-k,p,q}C_{n,p,q}(x,y)\right)\frac{t^n}{[n]_{p,q}!}. \tag{31}$$

Through comparison of the coefficients of both sides for Equation (31), we obtain the desired results immediately.

(ii) By applying a method similar to (i) in the generating function of the (p,q)-sine Bernoulli polynomials, we complete the proof of Theorem 4 (ii).

$\square$

Theorem 5. *For a nonnegative integer n, we derive*

$$(i) \quad [n]_{p,q}C_{n-1,p,q}(x,y) = \sum_{k=0}^{n}\begin{bmatrix}n\\k\end{bmatrix}_{p,q} p^{\binom{n-k}{2}}{}_{C}B_{k,p,q}(x,y) - {}_{C}B_{n,p,q}(x,y),$$

$$(ii) \quad [n]_{p,q}S_{n-1,p,q}(x,y) = \sum_{k=0}^{n}\begin{bmatrix}n\\k\end{bmatrix}_{p,q} p^{\binom{n-k}{2}}{}_{S}B_{k,p,q}(x,y) - {}_{S}B_{n,p,q}(x,y). \tag{32}$$

Proof.

(i) Suppose that $e_{p,q}(t) \neq 1$ in the generating function of the (p,q)-cosine Bernoulli polynomials. Then, we have

$$\sum_{n=0}^{\infty} {}_{C}B_{n,p,q}(x,y)\frac{t^n}{[n]_{p,q}!}(e_{p,q}(t) - 1) = te_{p,q}(tx)COS_{p,q}(ty). \tag{33}$$

We write the left-hand side of Equation (33) as follows:

$$\sum_{n=0}^{\infty} {}_{C}B_{n,p,q}(x,y)\frac{t^n}{[n]_{p,q}!}(e_{p,q}(t) - 1)$$

$$= \sum_{n=0}^{\infty} {}_{C}B_{n,p,q}(x,y)\frac{t^n}{[n]_{p,q}!}\left(\sum_{n=0}^{\infty} p^{\binom{n}{2}}\frac{t^n}{[n]_{p,q}!} - 1\right) \tag{34}$$

$$= \sum_{n=0}^{\infty}\left(\sum_{k=0}^{n}\begin{bmatrix}n\\k\end{bmatrix}_{p,q} p^{\binom{n-k}{2}}{}_{C}B_{k,p,q}(x,y) - {}_{C}B_{n,p,q}(x,y)\right)\frac{t^n}{[n]_{p,q}!},$$

and we transform the right-hand side into the following:

$$te_{p,q}(tx)COS_{p,q}(ty) = \sum_{n=0}^{\infty} C_{n,p,q}(x,y)\frac{t^{n+1}}{[n]_{p,q}!}$$

$$= \sum_{n=0}^{\infty} [n]_{p,q}C_{n-1,p,q}(x,y)\frac{t^n}{[n]_{p,q}!}. \tag{35}$$

Therefore, we obtain the following:

$$\sum_{k=0}^{n}\begin{bmatrix}n\\k\end{bmatrix}_{p,q} p^{\binom{n-k}{2}}{}_{C}B_{k,p,q}(x,y) - {}_{C}B_{n,p,q}(x,y) = [n]_{p,q}C_{n-1,p,q}(x,y). \tag{36}$$

By calculating the left-hand side of Equation (36), we investigate the required result.

(ii)   We do not include the proof of Theorem 5 (ii) because the proving process is similar to that of Theorem 5 (i).

$\square$

Corollary 2. *Setting $p = 1$ in Theorem 5, the following equations hold*

$$(i) \quad [n]_q C_{n-1,q}(x,y) = \sum_{k=0}^{n} \begin{bmatrix} n \\ k \end{bmatrix}_q {}_C B_{k,q}(x,y) - {}_C B_{n,q}(x,y)$$

$$(ii) \quad [n]_q S_{n-1,q}(x,y) = \sum_{k=0}^{n} \begin{bmatrix} n \\ k \end{bmatrix}_q {}_S B_{k,q}(x,y) - {}_S B_{n,q}(x,y),$$

$$(37)$$

where ${}_C B_{n,q}(x,y)$ represents the q-cosine Bernoulli polynomials and ${}_S B_{n,q}(x,y)$ denotes the q-sine Bernoulli polynomials.

Corollary 3. *Assigning $p = 1$ and $q \to 1$ in Theorem 5, the following holds:*

$$(i) \quad n C_{n-1}(x,y) = \sum_{k=0}^{n-1} \binom{n}{k} {}_C B_n(x,y)$$

$$(ii) \quad n S_{n-1}(x,y) = \sum_{k=0}^{n-1} \binom{n}{k} {}_S B_n(x,y),$$

$$(38)$$

where ${}_C B_n(x,y)$ represents the cosine Bernoulli polynomials and ${}_S B_n(x,y)$ represents the sine Bernoulli polynomials.

Theorem 6. *Let $|p/q| < 1$. Then, we have*

$$(i) \quad {}_C B_{n,p,q}(1,y) = \sum_{k=0}^{n} \begin{bmatrix} n \\ k \end{bmatrix}_{p,q} (-1)^{n-k} q^{\binom{n-k}{2}} \left([k]_{p,q} C_{k-1,p,q}(x,y) + {}_C B_{k,p,q}(x,y) \right) x^{n-k},$$

$$(ii) \quad {}_S B_{n,p,q}(1,y) = \sum_{k=0}^{n} \begin{bmatrix} n \\ k \end{bmatrix}_{p,q} (-1)^{n-k} q^{\binom{n-k}{2}} \left([k]_{p,q} S_{k-1,p,q}(x,y) + {}_S B_{k,p,q}(x,y) \right) x^{n-k}.$$

$$(39)$$

Proof.

(i)   If we put 1 instead of x in the generating function of the (p,q)-cosine Bernoulli polynomials, we find the following:

$$\sum_{n=0}^{\infty} {}_C B_{n,p,q}(1,y) \frac{t^n}{[n]_{p,q}!} = \frac{t}{e_{p,q}(t) - 1} \left(e_{p,q}(t) - 1 + 1 \right) COS_{p,q}(ty)$$

$$= t COS_{p,q}(ty) + \frac{t}{e_{p,q}(t) - 1} COS_{p,q}(ty).$$

$$(40)$$

Using a property of the (p,q)-exponential function, $e_{p,q}(x)E_{p,q}(-x) = 1$, in Equation (40), we obtain the following:

$$\sum_{n=0}^{\infty} {}_cB_{n,p,q}(1,y)\frac{t^n}{[n]_{p,q}!}$$
$$= \left(te_{p,q}(tx)COS_{p,q}(ty) + \frac{t}{e_{p,q}(t)-1}e_{p,q}(tx)COS_{p,q}(ty) \right) E_{p,q}(-tx)$$
$$= \sum_{n=0}^{\infty} \left([n]_{p,q}C_{n-1,p,q}(x,y) + {}_cB_{n,p,q}(x,y)\right) \frac{t^n}{[n]_{p,q}!} \sum_{n=0}^{\infty} q^{\binom{n}{2}}(-x)^n\frac{t^n}{[n]_{p,q}!}$$
$$= \sum_{n=0}^{\infty} \left(\sum_{k=0}^{n} \begin{bmatrix} n \\ k \end{bmatrix}_{p,q} (-1)^{n-k}q^{\binom{n-k}{2}}([k]_{p,q}C_{k-1,p,q}(x,y) + {}_cB_{k,p,q}(x,y))x^{n-k} \right)\frac{t^n}{[n]_{p,q}!},$$

(41)

and we immediately derive the results.

(ii)   By applying a similar process for proving (i) to the (p,q)-sine Bernoulli polynomials, we find Theorem 6 (ii).

$\square$

Corollary 4. *Setting $p = 1$ in Theorem 6, the following holds:*

$$(i) \quad {}_cB_{n,q}(1,y) = \sum_{k=0}^{n} \begin{bmatrix} n \\ k \end{bmatrix}_q (-1)^{n-k}q^{\binom{n-k}{2}} \left([k]_qC_{k-1,q}(x,y) + {}_cB_{k,q}(x,y)\right) x^{n-k}$$
$$(ii) \quad {}_sB_{n,q}(1,y) = \sum_{k=0}^{n} \begin{bmatrix} n \\ k \end{bmatrix}_q (-1)^{n-k}q^{\binom{n-k}{2}} \left([k]_qS_{k-1,q}(x,y) + {}_sB_{k,q}(x,y)\right) x^{n-k},$$

(42)

where ${}_cB_{n,q}(x,y)$ denotes the q-cosine Bernoulli polynomials and ${}_sB_{n,q}(x,y)$ denotes the q-sine Bernoulli polynomials.

Corollary 5. *Setting $p = 1$ and $q \to 1$ in Theorem 6, the following holds:*

$$(i) \quad {}_cB_n(1,y) = \sum_{k=0}^{n} \binom{n}{k}(-1)^{n-k}\left(kC_{k-1}(x,y) + {}_cB_k(x,y)\right) x^{n-k}$$
$$(ii) \quad {}_sB_n(1,y) = \sum_{k=0}^{n} \binom{n}{k}(-1)^{n-k}\left(kS_{k-1}(x,y) + {}_sB_k(x,y)\right) x^{n-k},$$

(43)

where ${}_cB_n(x,y)$ is the cosine Bernoulli polynomials and ${}_sB_n(x,y)$ is the sine Bernoulli polynomials.

Theorem 7. *For a nonnegative integer k and $|p/q| < 1$, we investigate*

$$B_{n,p,q}(x) = \sum_{k=0}^{\left[\frac{n}{2}\right]} \begin{bmatrix} n \\ 2k \end{bmatrix}_{p,q} (-1)^k p^{(2k-1)k}y^{2k}{}_cB_{n-k,p,q}(x,y)$$
$$+ \sum_{k=0}^{\left[\frac{n-1}{2}\right]} \begin{bmatrix} n \\ 2k+1 \end{bmatrix}_{p,q} (-1)^k p^{(2k+1)k}y^{2k+1}{}_sB_{n-(2k+1),p,q}(x,y),$$

(44)

where $B_{n,p,q}(x)$ is the (p,q)-Bernoulli polynomials and $[x]$ is the greatest integer not exceeding x.

Proof. In [9], we observe the power series of (p,q)-cosine and (p,q)-sine functions as follows:

$$cos_{p,q}(x) = \sum_{n=0}^{\infty} (-1)^n p^{(2n-1)n} \frac{x^{2n}}{[2n]_{p,q}!}, \quad sin_{p,q}(x) = \sum_{n=0}^{\infty} (-1)^n p^{(2n+1)n} \frac{x^{2n+1}}{[2n+1]_{p,q}!}. \tag{45}$$

Let us consider (p,q)-cosine Bernoulli polynomials. If we multiply (p,q)-cosine Bernoulli polynomials and the (p,q)-cosine function to determine the relationship between (p,q)-Bernoulli polynomials and, combined (p,q)-cosine Bernoulli polynomials, and (p,q)-sine Bernoulli polynomials, we have

$$\sum_{n=0}^{\infty} {}_C B_{n,p,q}(x,y) \frac{t^n}{[n]_{p,q}!} cos_{p,q}(ty) = \frac{t}{e_{p,q}(t)-1} e_{p,q}(tx) COS_{p,q}(ty) cos_{p,q}(ty). \tag{46}$$

The left-hand side of Equation (46) is transformed as

$$\begin{aligned}
&\sum_{n=0}^{\infty} {}_C B_{n,p,q}(x,y) \frac{t^n}{[n]_{p,q}!} cos_{p,q}(ty) \\
&= \sum_{n=0}^{\infty} {}_C B_{n,p,q}(x,y) \frac{t^n}{[n]_{p,q}!} \sum_{n=0}^{\infty} (-1)^n p^{(2n-1)n} y^{2n} \frac{t^{2n}}{[n]_{p,q}!} \\
&= \sum_{n=0}^{\infty} \left(\sum_{k=0}^{n} \begin{bmatrix} n+k \\ 2k \end{bmatrix}_{p,q} (-1)^k p^{(2k-1)k} y^{2k} {}_C B_{n-k,p,q}(x,y) \right) \frac{t^{n+k}}{[n+k]_{p,q}!} \\
&= \sum_{n=0}^{\infty} \left(\sum_{k=0}^{[\frac{n}{2}]} \begin{bmatrix} n \\ 2k \end{bmatrix}_{p,q} (-1)^k p^{(2k-1)k} y^{2k} {}_C B_{n-k,p,q}(x,y) \right) \frac{t^n}{[n]_{p,q}!}.
\end{aligned} \tag{47}$$

From Equations (46) and (47), we derive the following:

$$\begin{aligned}
&\sum_{n=0}^{\infty} \left(\sum_{k=0}^{[\frac{n}{2}]} \begin{bmatrix} n \\ 2k \end{bmatrix}_{p,q} (-1)^k p^{(2k-1)k} y^{2k} {}_C B_{n-k,p,q}(x,y) \right) \frac{t^n}{[n]_{p,q}!} \\
&= \frac{t}{e_{p,q}(t)-1} e_{p,q}(tx) COS_{p,q}(ty) cos_{p,q}(ty),
\end{aligned} \tag{48}$$

where $[x]$ is the greatest integer that does not exceed x.

From now on, let us consider the (p,q)-sine Bernoulli polynomials in a same manner of (p,q)-cosine Bernoulli polynomials. If we multiply ${}_S B_{n,p,q}(x,y)$ and $sin_{p,q}(ty)$, we obtain

$$\sum_{n=0}^{\infty} {}_S B_{n,p,q}(x,y) \frac{t^n}{[n]_{p,q}!} sin_{p,q}(ty) = \frac{t}{e_{p,q}(t)-1} e_{p,q}(tx) SIN_{p,q}(ty) sin_{p,q}(ty). \tag{49}$$

The left-hand side of Equation (49) can be changed as the following.

$$\begin{aligned}
&\sum_{n=0}^{\infty} {}_S B_{n,p,q}(x,y) \frac{t^n}{[n]_{p,q}!} sin_{p,q}(ty) \\
&= \sum_{n=0}^{\infty} {}_S B_{n,p,q}(x,y) \frac{t^n}{[n]_{p,q}!} \sum_{n=0}^{\infty} (-1)^n p^{(2n+1)n} y^{2n+1} \frac{t^{2n+1}}{[n]_{p,q}!} \\
&= \sum_{n=0}^{\infty} \left(\sum_{k=0}^{[\frac{n-1}{2}]} \begin{bmatrix} n \\ 2k+1 \end{bmatrix}_{p,q} (-1)^k p^{(2k+1)k} y^{2k+1} {}_S B_{n-(2k+1),p,q}(x,y) \right) \frac{t^n}{[n]_{p,q}!}.
\end{aligned} \tag{50}$$

From Equations (49) and (50), we have the following:

$$\sum_{n=0}^{\infty}\left(\sum_{k=0}^{[\frac{n-1}{2}]}\begin{bmatrix}n\\2k+1\end{bmatrix}_{p,q}(-1)^k p^{(2k+1)k}y^{2k+1}{}_{S}B_{n-(2k+1),p,q}(x,y)\right)\frac{t^n}{[n]_{p,q}!}$$
$$=\frac{t}{e_{p,q}(t)-1}e_{p,q}(tx)SIN_{p,q}(ty)sin_{p,q}(ty),$$

(51)

where $[x]$ is the greatest integer that does not exceed x.

Here, we recall that

$$\left(COS_{p,q}(x)cos_{p,q}(x)+SIN_{p,q}(x)sin_{p,q}(x)\right)=1.$$

(52)

Using Equations (48) and (51) and applying the property of (p,q)-trigonometric functions, we find (p,q)-Bernoulli polynomials as follows:

$$\sum_{n=0}^{\infty}\left(\sum_{k=0}^{[\frac{n}{2}]}\begin{bmatrix}n\\2k\end{bmatrix}_{p,q}(-1)^k p^{(2k-1)k}y^{2k}{}_{C}B_{n-k,p,q}(x,y)\right)\frac{t^n}{[n]_{p,q}!}$$
$$+\sum_{n=0}^{\infty}\left(\sum_{k=0}^{[\frac{n-1}{2}]}\begin{bmatrix}n\\2k+1\end{bmatrix}_{p,q}(-1)^k p^{(2k+1)k}y^{2k+1}{}_{S}B_{n-(2k+1),p,q}(x,y)\right)\frac{t^n}{[n]_{p,q}!}$$
$$=\frac{t}{e_{p,q}(t)-1}e_{p,q}(tx)\left(COS_{p,q}(ty)cos_{p,q}(ty)+SIN_{p,q}(ty)sin_{p,q}(ty)\right)$$
$$=\sum_{n=0}^{\infty}B_{n,p,q}(x)\frac{t^n}{[n]_{p,q}!},$$

(53)

where $B_{n,p,q}(x)$ is the (p,q)-Bernoulli polynomials.

By comparing the coefficients of both sides of t^n, we produce the desired result. □

Corollary 6. *Setting $p=1$ in Theorem 7, the following holds:*

$$B_{n,q}(x)=\sum_{k=0}^{[\frac{n}{2}]}\begin{bmatrix}n\\2k\end{bmatrix}_{q}(-1)^k y^{2k}{}_{C}B_{n-k,q}(x,y)$$
$$+\sum_{k=0}^{[\frac{n-1}{2}]}\begin{bmatrix}n\\2k+1\end{bmatrix}_{q}(-1)^k y^{2k+1}{}_{S}B_{n-(2k+1),q}(x,y),$$

(54)

where $B_{n,q}(x)$ is the q-Bernoulli polynomials, ${}_{C}B_{n,q}(x,y)$ denote the q-cosine Bernoulli polynomials, and ${}_{S}B_{n,q}(x)$ denote the q-sine Bernoulli polynomials.

Corollary 7. *Setting $y=1$ in Theorem 7, one holds:*

$$B_{n,p,q}(x)=\sum_{k=0}^{[\frac{n}{2}]}\begin{bmatrix}n\\2k\end{bmatrix}_{p,q}(-1)^k p^{(2k-1)k}{}_{C}B_{n-k,p,q}(x,1)$$
$$+\sum_{k=0}^{[\frac{n-1}{2}]}\begin{bmatrix}n\\2k+1\end{bmatrix}_{p,q}(-1)^k p^{(2k+1)k}{}_{S}B_{n-(2k+1),p,q}(x,1),$$

(55)

where $B_{n,p,q}(x)$ is the (p,q)-Bernoulli polynomials and $[x]$ is the greatest integers that does not exceed x.

Theorem 8. *For a nonnegative integer k and $|p/q| < 1$, we derive*

$$
\sum_{k=0}^{\left[\frac{n-1}{2}\right]} \begin{bmatrix} n \\ 2k+1 \end{bmatrix}_{p,q} (-1)^k p^{(2k+1)k} y^{2k+1}\, {}_C B_{n-(2k+1),p,q}(x,y)
$$
$$
= \sum_{k=0}^{\left[\frac{n}{2}\right]} \begin{bmatrix} n \\ 2k \end{bmatrix}_{p,q} (-1)^k p^{(2k-1)k} y^{2k}\, {}_S B_{n-k,p,q}(x,y),
$$
(56)

where $[x]$ is the greatest integer not exceeding x.

Proof. If we multiply ${}_C B_{n,p,q}(x,y)$ and $\sin_{p,q}(ty)$, then we find

$$
\sum_{n=0}^{\infty} {}_C B_{n,p,q}(x,y) \frac{t^n}{[n]_{p,q}!} \sin_{p,q}(ty) = \frac{t}{e_{p,q}(t)-1} e_{p,q}(tx) COS_{p,q}(ty) \sin_{p,q}(ty),
$$
(57)

and the left-hand side of Equation (57) can be transformed as

$$
\sum_{n=0}^{\infty} {}_C B_{n,p,q}(x,y) \frac{t^n}{[n]_{p,q}!} \sin_{p,q}(ty)
$$
$$
= \sum_{n=0}^{\infty} {}_C B_{n,p,q}(x,y) \frac{t^n}{[n]_{p,q}!} \sum_{n=0}^{\infty} (-1)^n p^{(2n+1)n} y^{2n+1} \frac{t^{2n+1}}{[2n+1]_{p,q}!}
$$
$$
= \sum_{n=0}^{\infty} \left(\sum_{k=0}^{\left[\frac{n-1}{2}\right]} \begin{bmatrix} n \\ 2k+1 \end{bmatrix}_{p,q} (-1)^k p^{(2k+1)k} y^{2k+1}\, {}_C B_{n-(2k+1),p,q}(x,y) \right) \frac{t^n}{[n]_{p,q}!}.
$$
(58)

Similarly, we multiply the (p,q)-sine Bernoulli polynomials and (p,q)-cosine function as the follows:

$$
\sum_{n=0}^{\infty} {}_S B_{n,p,q}(x,y) \frac{t^n}{[n]_{p,q}!} \cos_{p,q}(ty) = \frac{t}{e_{p,q}(t)-1} e_{p,q}(tx) SIN_{p,q}(ty) \cos_{p,q}(ty).
$$
(59)

The left-hand side of Equation (59) can be changed as

$$
\sum_{n=0}^{\infty} {}_S B_{n,p,q}(x,y) \frac{t^n}{[n]_{p,q}!} \cos_{p,q}(ty)
$$
$$
= \sum_{n=0}^{\infty} {}_S B_{n,p,q}(x,y) \frac{t^n}{[n]_{p,q}!} \sum_{n=0}^{\infty} (-1)^n p^{(2n-1)n} y^{2n} \frac{t^{2n}}{[2n]_{p,q}!}
$$
$$
= \sum_{n=0}^{\infty} \left(\sum_{k=0}^{\left[\frac{n-1}{2}\right]} \begin{bmatrix} n \\ 2k \end{bmatrix}_{p,q} (-1)^k p^{(2k-1)k} y^{2k}\, {}_S B_{n-k,p,q}(x,y) \right) \frac{t^n}{[n]_{p,q}!}.
$$
(60)

In here, we recall that $\sin_{p,q}(x) COS_{p,q}(x) = \cos_{p,q}(x) SIN_{p,q}(x)$. From Equations (58) and (60), and the above property of (p,q)-trigonometric functions, we investigate

$$
\sum_{n=0}^{\infty} \left(\sum_{k=0}^{\left[\frac{n-1}{2}\right]} \begin{bmatrix} n \\ 2k+1 \end{bmatrix}_{p,q} (-1)^k p^{(2k+1)k} y^{2k+1}\, {}_C B_{n-(2k+1),p,q}(x,y) \right) \frac{t^n}{[n]_{p,q}!}
$$
$$
- \sum_{n=0}^{\infty} \left(\sum_{k=0}^{\left[\frac{n}{2}\right]} \begin{bmatrix} n \\ 2k \end{bmatrix}_{p,q} (-1)^k p^{(2k-1)k} y^{2k}\, {}_S B_{n-k,p,q}(x,y) \right) \frac{t^n}{[n]_{p,q}!}
$$
$$
= \frac{t}{e_{p,q}(t)-1} e_{p,q}(tx) \left(COS_{p,q}(ty) \sin_{p,q}(ty) - SIN_{p,q}(ty) \cos_{p,q}(ty) \right)
$$
(61)

From Equation (61), we complete the proof of Theorem 8. $\quad\square$

Corollary 8. *Putting $y = 1$ in Theorem 8, we have the following:*

$$
\sum_{k=0}^{[\frac{n-1}{2}]} \begin{bmatrix} n \\ 2k+1 \end{bmatrix}_{p,q} (-1)^k p^{(2k+1)k} {}_c B_{n-(2k+1),p,q}(x,1)
$$
$$
= \sum_{k=0}^{[\frac{n}{2}]} \begin{bmatrix} n \\ 2k \end{bmatrix}_{p,q} (-1)^k p^{(2k-1)k} {}_s B_{n-k,p,q}(x,1),
\tag{62}
$$

where $[x]$ is the greatest integer not exceeding x.

Corollary 9. *Setting $p = 1$ in Theorem 8, the following holds:*

$$
\sum_{k=0}^{[\frac{n-1}{2}]} \begin{bmatrix} n \\ 2k+1 \end{bmatrix}_{q} (-1)^k y^{2k+1} {}_c B_{n-(2k+1),q}(x,y) = \sum_{k=0}^{[\frac{n}{2}]} \begin{bmatrix} n \\ 2k \end{bmatrix}_{q} (-1)^k y^{2k} {}_s B_{n-k,q}(x,y),
\tag{63}
$$

where ${}_c B_{n,q}(x,y)$ is the q-cosine Bernoulli polynomials and ${}_s B_{n,q}(x,y)$ is the q-sine Bernoulli polynomials.

Corollary 10. *Let $p = 1$ and $q \to 1$ in Theorem 8. Then one holds*

$$
\sum_{k=0}^{[\frac{n-1}{2}]} \binom{n}{2k+1} (-1)^k y^{2k+1} {}_c B_{n-(2k+1)}(x,y) = \sum_{k=0}^{[\frac{n}{2}]} \binom{n}{2k} (-1)^k y^{2k} {}_s B_{n-k}(x,y),
\tag{64}
$$

where ${}_c B_n(x,y)$ is the cosine Bernoulli polynomials and ${}_s B_n(x,y)$ is the sine Bernoulli polynomials.

4. Several Symmetric Properties of the (p,q)-Cosine and (p,q)-Sine Bernoulli Polynomials

In this section, we point out several symmetric identities of the (p,q)-cosine and (p,q)-Bernoulli polynomials. Using various forms that are made by a and b, we obtain a few desired results regarding the (p,q)-cosine and (p,q)-sine Bernoulli polynomials. Moreover, we discover other relations of different Bernoulli polynomials by considering certain conditions in theorems. We also find the symmetric structure of the approximate roots based on the symmetric polynomials.

Theorem 9. *Let a and b be nonzero. Then, we obtain*

$$
(i) \quad \sum_{k=0}^{n} \begin{bmatrix} n \\ k \end{bmatrix}_{p,q} a^{n-k-1} b^{k-1} {}_c B_{n-k,p,q}\left(\frac{x}{a},\frac{y}{a}\right) {}_c B_{k,p,q}\left(\frac{X}{b},\frac{Y}{b}\right)
$$
$$
= \sum_{k=0}^{n} \begin{bmatrix} n \\ k \end{bmatrix}_{p,q} b^{n-k-1} a^{k-1} {}_c B_{n-k,p,q}\left(\frac{x}{b},\frac{y}{b}\right) {}_c B_{k,p,q}\left(\frac{X}{a},\frac{Y}{a}\right),
$$
$$
(ii) \quad \sum_{k=0}^{n} \begin{bmatrix} n \\ k \end{bmatrix}_{p,q} a^{n-k-1} b^{k-1} {}_s B_{n-k,p,q}\left(\frac{x}{a},\frac{y}{a}\right) {}_s B_{k,p,q}\left(\frac{X}{b},\frac{Y}{b}\right)
$$
$$
= \sum_{k=0}^{n} \begin{bmatrix} n \\ k \end{bmatrix}_{p,q} b^{n-k-1} a^{k-1} {}_s B_{n-k,p,q}\left(\frac{x}{b},\frac{y}{b}\right) {}_s B_{k,p,q}\left(\frac{X}{a},\frac{Y}{a}\right).
\tag{65}
$$

Proof.

(i)   We consider form A as follows:

$$A := \frac{t^2 e_{p,q}(tx) e_{p,q}(tX) COS_{p,q}(ty) COS_{p,q}(tY)}{\left(e_{p,q}(at) - 1\right)\left(e_{p,q}(bt) - 1\right)} \tag{66}$$

From form A, we find

$$
\begin{aligned}
A &= \frac{t}{e_{p,q}(at) - 1} e_{p,q}(tx) COS_{p,q}(ty) \frac{t}{e_{p,q}(bt) - 1} e_{p,q}(tX) COS_{p,q}(tY) \\
&= \sum_{n=0}^{\infty} a^{n-1} {}_C B_{n,p,q}\left(\frac{x}{a}, \frac{y}{a}\right) \frac{t^n}{[n]_{p,q}!} \sum_{n=0}^{\infty} b^{n-1} {}_C B_{n,p,q}\left(\frac{X}{b}, \frac{Y}{b}\right) \frac{t^n}{[n]_{p,q}!} \\
&= \sum_{n=0}^{\infty} \left(\sum_{k=0}^{n} \begin{bmatrix} n \\ k \end{bmatrix}_{p,q} a^{n-k-1} b^{k-1} {}_C B_{n-k,p,q}\left(\frac{x}{a}, \frac{y}{a}\right) {}_C B_{k,p,q}\left(\frac{X}{b}, \frac{Y}{b}\right) \right) \frac{t^n}{[n]_{p,q}!},
\end{aligned}
\tag{67}
$$

and form A of Equation (66) can be transformed into the following:

$$
\begin{aligned}
A &= \frac{t}{e_{p,q}(bt) - 1} e_{p,q}(tx) COS_{p,q}(ty) \frac{t}{e_{p,q}(at) - 1} e_{p,q}(tX) COS_{p,q}(tY) \\
&= \sum_{n=0}^{\infty} \left(\sum_{k=0}^{n} \begin{bmatrix} n \\ k \end{bmatrix}_{p,q} b^{n-k-1} a^{k-1} {}_C B_{n-k,p,q}\left(\frac{x}{b}, \frac{y}{b}\right) {}_C B_{k,p,q}\left(\frac{X}{a}, \frac{Y}{a}\right) \right) \frac{t^n}{[n]_{p,q}!}.
\end{aligned}
\tag{68}
$$

Using the comparison of coefficients in Equations (67) and (68), we find the desired result.

(ii)   If we assume form B as follows:

$$B := \frac{t^2 e_{p,q}(tx) e_{p,q}(tX) SIN_{p,q}(ty) SIN_{p,q}(tY)}{\left(e_{p,q}(at) - 1\right)\left(e_{p,q}(bt) - 1\right)}, \tag{69}$$

then, we find Theorem 9 (ii) in the same manner.

$\square$

Corollary 11. *Setting $a = 1$ in Theorem 9, the following holds:*

$$
\begin{aligned}
(i)\quad & \sum_{k=0}^{n} \begin{bmatrix} n \\ k \end{bmatrix}_{p,q} b^{k-1} {}_C B_{n-k,p,q}(x,y) \, {}_C B_{k,p,q}\left(\frac{X}{b}, \frac{Y}{b}\right) \\
&= \sum_{k=0}^{n} \begin{bmatrix} n \\ k \end{bmatrix}_{p,q} b^{n-k-1} {}_C B_{n-k,p,q}\left(\frac{x}{b}, \frac{y}{b}\right) {}_C B_{k,p,q}(X,Y), \\
(ii)\quad & \sum_{k=0}^{n} \begin{bmatrix} n \\ k \end{bmatrix}_{p,q} b^{k-1} {}_S B_{n-k,p,q}(x,y) \, {}_S B_{k,p,q}\left(\frac{X}{b}, \frac{Y}{b}\right) \\
&= \sum_{k=0}^{n} \begin{bmatrix} n \\ k \end{bmatrix}_{p,q} b^{n-k-1} {}_S B_{n-k,p,q}\left(\frac{x}{b}, \frac{y}{b}\right) {}_S B_{k,p,q}(X,Y).
\end{aligned}
\tag{70}
$$

Corollary 12. *If $p = 1$ in Theorem 9, then we have*

$$
(i) \quad \sum_{k=0}^{n} \begin{bmatrix} n \\ k \end{bmatrix}_q a^{n-k-1} b^{k-1}\, {}_C B_{n-k,q}\left(\frac{x}{a}, \frac{y}{a}\right) {}_C B_{k,q}\left(\frac{X}{b}, \frac{Y}{b}\right)
$$

$$
= \sum_{k=0}^{n} \begin{bmatrix} n \\ k \end{bmatrix}_q b^{n-k-1} a^{k-1}\, {}_C B_{n-k,q}\left(\frac{x}{b}, \frac{y}{b}\right) {}_C B_{k,q}\left(\frac{X}{a}, \frac{Y}{a}\right),
$$

$$
(ii) \quad \sum_{k=0}^{n} \begin{bmatrix} n \\ k \end{bmatrix}_q a^{n-k-1} b^{k-1}\, {}_S B_{n-k,q}\left(\frac{x}{a}, \frac{y}{a}\right) {}_S B_{k,q}\left(\frac{X}{b}, \frac{Y}{b}\right)
$$

$$
= \sum_{k=0}^{n} \begin{bmatrix} n \\ k \end{bmatrix}_q b^{n-k-1} a^{k-1}\, {}_S B_{n-k,q}\left(\frac{x}{b}, \frac{y}{b}\right) {}_S B_{k,q}\left(\frac{X}{a}, \frac{Y}{a}\right), \tag{71}
$$

where ${}_C B_{n,q}(x,y)$ denotes the q-cosine Bernoulli polynomials and ${}_S B_{n,q}(x,y)$ denotes the q-sine Bernoulli polynomials.

Corollary 13. *Putting $p = 1$ and $q \to 1$, one holds:*

$$
(i) \quad \sum_{k=0}^{n} \binom{n}{k} a^{n-k-1} b^{k-1}\, {}_C B_{n-k}\left(\frac{x}{a}, \frac{y}{a}\right) {}_C B_{k}\left(\frac{X}{b}, \frac{Y}{b}\right)
$$

$$
= \sum_{k=0}^{n} \binom{n}{k} b^{n-k-1} a^{k-1}\, {}_C B_{n-k}\left(\frac{x}{b}, \frac{y}{b}\right) {}_C B_{k}\left(\frac{X}{a}, \frac{Y}{a}\right),
$$

$$
(ii) \quad \sum_{k=0}^{n} \binom{n}{k} a^{n-k-1} b^{k-1}\, {}_S B_{n-k}\left(\frac{x}{a}, \frac{y}{a}\right) {}_S B_{k}\left(\frac{X}{b}, \frac{Y}{b}\right)
$$

$$
= \sum_{k=0}^{n} \binom{n}{k} b^{n-k-1} a^{k-1}\, {}_S B_{n-k}\left(\frac{x}{b}, \frac{y}{b}\right) {}_S B_{k}\left(\frac{X}{a}, \frac{Y}{a}\right), \tag{72}
$$

where ${}_C B_n(x,y)$ is the cosine Bernoulli polynomials and ${}_S B_n(x,y)$ is the sine Bernoulli polynomials.

Theorem 9 is a basic symmetric property of (p,q)-cosine and (p,q)-sine Bernoulli polynomials. We aim to find several symmetric properties by mixing (p,q)-cosine and (p,q)-sine Bernoulli polynomials.

Theorem 10. *For nonzero integers a and b, we have*

$$
\sum_{k=0}^{n} \begin{bmatrix} n \\ k \end{bmatrix}_{p,q} a^{n-k-1} b^{k-1}\, {}_C B_{n-k,p,q}\left(\frac{x}{a}, \frac{y}{a}\right) {}_S B_{k,p,q}\left(\frac{X}{b}, \frac{Y}{b}\right)
$$

$$
= \sum_{k=0}^{n} \begin{bmatrix} n \\ k \end{bmatrix}_{p,q} b^{n-k-1} a^{k-1}\, {}_C B_{n-k,p,q}\left(\frac{x}{b}, \frac{y}{b}\right) {}_S B_{k,p,q}\left(\frac{X}{a}, \frac{Y}{a}\right). \tag{73}
$$

Proof. We assume form C by mixing the (p,q)-cosine function with the (p,q)-sine function, such as the following:

$$
C := \frac{t^2 e_{p,q}(tx) e_{p,q}(tX) COS_{p,q}(ty) SIN_{p,q}(tY)}{(e_{p,q}(at) - 1)(e_{p,q}(bt) - 1)}. \tag{74}
$$

Form C of the above equation can be changed into

$$C = \frac{t}{e_{p,q}(at)-1}e_{p,q}(tx)COS_{p,q}(ty)\frac{t}{e_{p,q}(bt)-1}e_{p,q}(tX)SIN_{p,q}(tY)$$

$$= \sum_{n=0}^{\infty} a^{n-1}{}_C B_{n,p,q}\left(\frac{x}{a},\frac{y}{a}\right)\frac{t^n}{[n]_{p,q}!}\sum_{n=0}^{\infty} b^{n-1}{}_S B_{n,p,q}\left(\frac{X}{b},\frac{Y}{b}\right)\frac{t^n}{[n]_{p,q}!} \tag{75}$$

$$= \sum_{n=0}^{\infty}\left(\sum_{k=0}^{n}\begin{bmatrix}n\\k\end{bmatrix}_{p,q} a^{n-k-1}b^{k-1}{}_C B_{n-k,p,q}\left(\frac{x}{a},\frac{y}{a}\right){}_S B_{k,p,q}\left(\frac{X}{b},\frac{Y}{b}\right)\right)\frac{t^n}{[n]_{p,q}!},$$

or, equivalently:

$$C = \frac{t}{e_{p,q}(bt)-1}e_{p,q}(tx)COS_{p,q}(ty)\frac{t}{e_{p,q}(at)-1}e_{p,q}(tX)SIN_{p,q}(tY)$$

$$\tag{76}$$

$$= \sum_{n=0}^{\infty}\left(\sum_{k=0}^{n}\begin{bmatrix}n\\k\end{bmatrix}_{p,q} b^{n-k-1}a^{k-1}{}_C B_{n-k,p,q}\left(\frac{x}{b},\frac{y}{b}\right){}_S B_{k,p,q}\left(\frac{X}{a},\frac{Y}{a}\right)\right)\frac{t^n}{[n]_{p,q}!}.$$

By comparing transformed Equations (75) and (76), we determine the result of Theorem 10. $\square$

Corollary 14. *If $a = 1$ in Theorem 10, then we find*

$$\sum_{k=0}^{n}\begin{bmatrix}n\\k\end{bmatrix}_{p,q} b^{k-1}{}_C B_{n-k,p,q}(x,y){}_S B_{k,p,q}\left(\frac{X}{b},\frac{Y}{b}\right)$$

$$\tag{77}$$

$$= \sum_{k=0}^{n}\begin{bmatrix}n\\k\end{bmatrix}_{p,q} b^{n-k-1}{}_C B_{n-k,p,q}\left(\frac{x}{b},\frac{y}{b}\right){}_S B_{k,p,q}(X,Y).$$

Corollary 15. *Setting $p = 1$ in Theorem 10, one holds:*

$$\sum_{k=0}^{n}\begin{bmatrix}n\\k\end{bmatrix}_{q} a^{n-k-1}b^{k-1}{}_C B_{n-k,q}\left(\frac{x}{a},\frac{y}{a}\right){}_S B_{k,q}\left(\frac{X}{b},\frac{Y}{b}\right)$$

$$\tag{78}$$

$$= \sum_{k=0}^{n}\begin{bmatrix}n\\k\end{bmatrix}_{q} b^{n-k-1}a^{k-1}{}_C B_{n-k,q}\left(\frac{x}{b},\frac{y}{b}\right){}_S B_{k,q}\left(\frac{X}{a},\frac{Y}{a}\right),$$

where ${}_C B_{n,q}(x,y)$ is the q-cosine Bernoulli polynomials and ${}_S B_{n,q}(x,y)$ is the q-sine Bernoulli polynomials.

Corollary 16. *Assigning $p = 1$ and $q \to 1$ in Theorem 10, the following holds:*

$$\sum_{k=0}^{n}\binom{n}{k} a^{n-k-1}b^{k-1}{}_C B_{n-k}\left(\frac{x}{a},\frac{y}{a}\right){}_S B_{k}\left(\frac{X}{b},\frac{Y}{b}\right)$$

$$\tag{79}$$

$$= \sum_{k=0}^{n}\binom{n}{k} b^{n-k-1}a^{k-1}{}_C B_{n-k}\left(\frac{x}{b},\frac{y}{b}\right){}_S B_{k}\left(\frac{X}{a},\frac{Y}{a}\right),$$

where ${}_C B_n(x,y)$ is the cosine Bernoulli polynomials and ${}_S B_n(x,y)$ is the sine Bernoulli polynomials.

Theorem 11. *Let a and b be nonzero integers. Then, we derive*

$$
\sum_{k=0}^{n} \begin{bmatrix} n \\ k \end{bmatrix}_{p,q} a^{n-k-1} b^{k-1}\, {}_C B_{n-k,p,q}\left(bx, \frac{y}{a}\right) {}_S B_{k,p,q}\left(aX, \frac{Y}{b}\right)
$$
$$
= \sum_{k=0}^{n} \begin{bmatrix} n \\ k \end{bmatrix}_{p,q} b^{n-k-1} a^{k-1}\, {}_C B_{n-k,p,q}\left(ax, \frac{y}{b}\right) {}_S B_{k,p,q}\left(bX, \frac{Y}{a}\right).
$$

(80)

Proof. Let us consider form D containing a and b in the (p,q)-exponential functions as

$$
D := \frac{t^2 e_{p,q}(abtx) e_{p,q}(abtX) COS_{p,q}(ty) SIN_{p,q}(tY)}{\left(e_{p,q}(at)-1\right)\left(e_{p,q}(bt)-1\right)}.
$$

(81)

From the above form D, we can obtain

$$
D = \frac{t}{e_{p,q}(at)-1} e_{p,q}(abtx) COS_{p,q}(ty) \frac{t}{e_{p,q}(bt)-1} e_{p,q}(abtX) SIN_{p,q}(tY)
$$
$$
= \sum_{n=0}^{\infty} a^{n-1}\, {}_C B_{n,p,q}\left(bx, \frac{y}{a}\right) \frac{t^n}{[n]_{p,q}!} \sum_{n=0}^{\infty} b^{n-1}\, {}_S B_{n,p,q}\left(aX, \frac{Y}{b}\right) \frac{t^n}{[n]_{p,q}!}
$$
$$
= \sum_{n=0}^{\infty} \left(\sum_{k=0}^{n} \begin{bmatrix} n \\ k \end{bmatrix}_{p,q} a^{n-k-1} b^{k-1}\, {}_C B_{n-k,p,q}\left(bx, \frac{y}{a}\right) {}_S B_{k,p,q}\left(aX, \frac{Y}{b}\right) \right) \frac{t^n}{[n]_{p,q}!},
$$

(82)

and

$$
D = \frac{t}{e_{p,q}(bt)-1} e_{p,q}(abtx) COS_{p,q}(ty) \frac{t}{e_{p,q}(at)-1} e_{p,q}(abtX) SIN_{p,q}(tY)
$$
$$
= \sum_{n=0}^{\infty} \left(\sum_{k=0}^{n} \begin{bmatrix} n \\ k \end{bmatrix}_{p,q} b^{n-k-1} a^{k-1}\, {}_C B_{n-k,p,q}\left(ax, \frac{y}{b}\right) {}_S B_{k,p,q}\left(bX, \frac{Y}{a}\right) \right) \frac{t^n}{[n]_{p,q}!}.
$$

(83)

By observing Equations (82) and (83) which are made by form D, we prove Theorem 11. $\square$

Corollary 17. *Setting $a = 1$ in Theorem 11, the following holds:*

$$
\sum_{k=0}^{n} \begin{bmatrix} n \\ k \end{bmatrix}_{p,q} b^{k-1}\, {}_C B_{n-k,p,q}(bx, y)\, {}_S B_{k,p,q}\left(X, \frac{Y}{b}\right)
$$
$$
= \sum_{k=0}^{n} \begin{bmatrix} n \\ k \end{bmatrix}_{p,q} b^{n-k-1}\, {}_C B_{n-k,p,q}\left(x, \frac{y}{b}\right) {}_S B_{k,p,q}(bX, Y).
$$

(84)

Corollary 18. *If $p = 1$ in Theorem 11, then we obtain*

$$
\sum_{k=0}^{n} \begin{bmatrix} n \\ k \end{bmatrix}_{q} a^{n-k-1} b^{k-1}\, {}_C B_{n-k,q}\left(bx, \frac{y}{a}\right) {}_S B_{k,q}\left(aX, \frac{Y}{b}\right)
$$
$$
= \sum_{k=0}^{n} \begin{bmatrix} n \\ k \end{bmatrix}_{q} b^{n-k-1} a^{k-1}\, {}_C B_{n-k,q}\left(ax, \frac{y}{b}\right) {}_S B_{k,q}\left(bX, \frac{Y}{a}\right),
$$

(85)

where ${}_C B_{n,q}(x, y)$ is the q-cosine Bernoulli polynomials and ${}_S B_{n,q}(x, y)$ is the q-sine Bernoulli polynomials.

Corollary 19. *Let $p = 1$ and $q \to 1$ in Theorem 11. Then one holds*

$$\sum_{k=0}^{n} \binom{n}{k} a^{n-k-1} b^{k-1}{}_C B_{n-k}\left(bx, \frac{y}{a}\right) {}_S B_k\left(aX, \frac{Y}{b}\right)$$

$$= \sum_{k=0}^{n} \binom{n}{k} b^{n-k-1} a^{k-1}{}_C B_{n-k}\left(ax, \frac{y}{b}\right) {}_S B_k\left(bX, \frac{Y}{a}\right), \tag{86}$$

where ${}_C B_n(x, y)$ is the cosine Bernoulli polynomials and ${}_S B_n(x, y)$ is the sine Bernoulli polynomials.

Next, we investigate the structure of approximate roots in (p, q)-cosine and (p, q)-sine Bernoulli polynomials. Based on the theorems above, (p, q)-cosine and (p, q)-sine Bernoulli polynomials have symmetric properties. Thus, we assume that the approximate roots of (p, q)-cosine and (p, q)-sine Bernoulli polynomials also have symmetric properties as well. We aim to identify the stacking structure of the roots from the specific (p, q)-cosine and (p, q)-sine Bernoulli polynomials found in Section 3.

First, the structure of approximate roots in the (p, q)-cosine Bernoulli polynomials is illustrated in Figure 1 when $y = 5$, $q = 0.9$, and the value of p changes. Figure 1 reveals the pattern of the roots in the (p, q)-cosine Bernoulli polynomials when $p = 0.5$. In addition, the approximate roots appear when n changes from 1 to 30. The red points become closer together when n is 30 and n becomes smaller as illustrated by the blue points. Based on the graphs with real and imaginary axes, the (p, q)-cosine Bernoulli polynomials are symmetric.

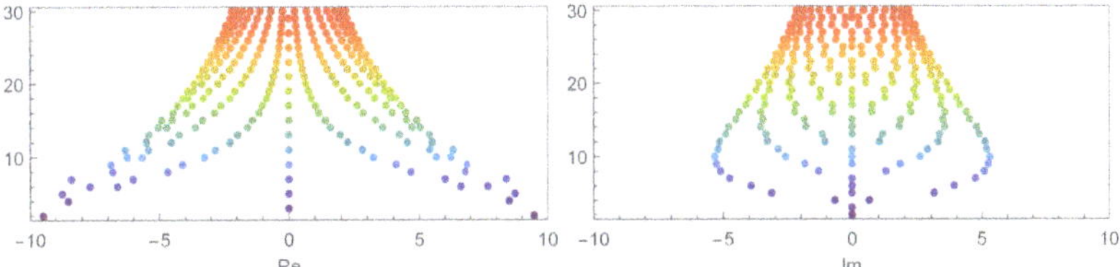

Figure 1. Stacking structure of approximate roots in the (p, q)-cosine Bernoulli polynomials when $p = 0.5, q = 0.9$, and $y = 5$.

Here, we aim to confirm that changes in the value of the (p, q)-cosine Bernoulli polynomials changes the structure of the approximate roots as the value changes. The structure of the approximate roots in polynomials when $p = 1$ and q changes, can be found in the q-cosine Bernoulli polynomials (see [11]).

Figure 2 below illustrates the stacking structure of the approximate roots of the (p, q)-cosine Bernoulli polynomials fixed at $p = 0.1, q = 0.5$ and $y = 5$ when $1 \leq n \leq 30$. Compared with Figure 1, Figure 2 displays a wider distribution of the approximate roots. The range of the left picture in Figure 1 is $-15 < \mathrm{Re}\, x < 15$ and the range of the left picture in Figure 2 is $-50 < \mathrm{Re}\, x < 50$. The structure of the approximate roots of $p = 0.1$ when $n = 30$ is wider on the real axis compared to when $p = 0.5$. The right-hand graphs in Figures 1 and 2 also reveal the same distribution. In addition, as n increases, the structure of the approximate roots appears symmetric.

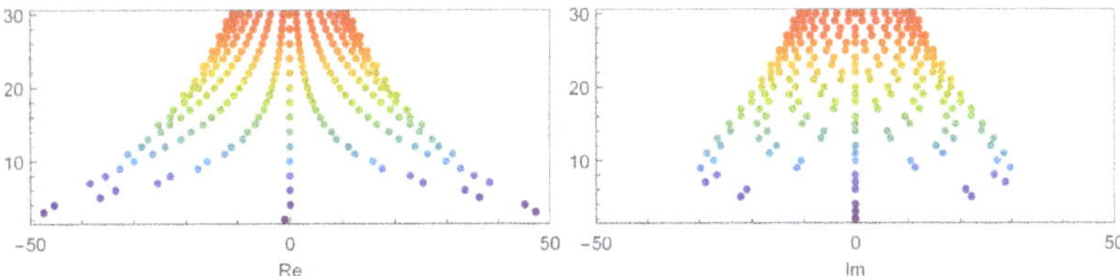

Figure 2. Stacking structure of approximate roots in the (p, q)-cosine Bernoulli polynomials when $p = 0.1, q = 0.9$, and $y = 5$.

Next, we examine the stacking structure of the approximate roots in the (p,q)-sine Bernoulli polynomials. The conditions are confirmed by equating them to the conditions of the (p,q)-cosine Bernoulli polynomials. The stacking structure of the approximate roots of the (p,q)-sine Bernoulli polynomials when $p = 0.5, q = 0.9$, and $y = 5$ can be checked in Figure 3. At $1 \leq n \leq 30$, the distribution range of the approximate roots appears wider in the values on the real axis than in the imaginary axis, as shown in the left picture in Figure 3. Figure 3 reveals that, as the value of n becomes larger, the approximate roots become more symmetric, and the approximate form approaches a circular shape, including the origin.

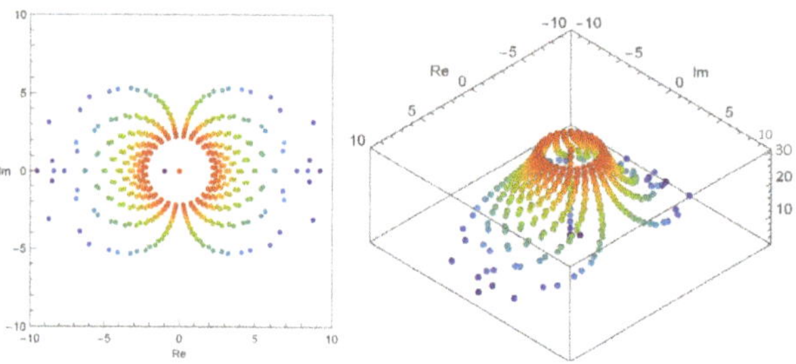

Figure 3. Stacking structure of approximate roots in the (p,q)-sine Bernoulli polynomials when $p = 0.5, q = 0.9$, and $y = 5$ in 3D.

When we change the value of p, the structure of the approximate roots of the (p,q)-sine Bernoulli polynomials when $p = 0.1$ under the same conditions as in Figure 3 is presented in Figure 4. In comparison with Figure 3, the area of the real and the imaginary axes in Figure 4 is greater, and the approximate roots have a wider distribution than observed in Figure 3. This property is common in the approximate roots of the (p,q)-cosine and (p,q)-sine Bernoulli polynomials. This indicates that, as the value of p decreases, the approximate roots of the (p,q)-cosine and (p,q)-sine Bernoulli polynomials spread wider. In addition, as displayed in Figure 4, the structure of the approximate roots of the (p,q)-sine Bernoulli polynomials is symmetric as the value of n increases.

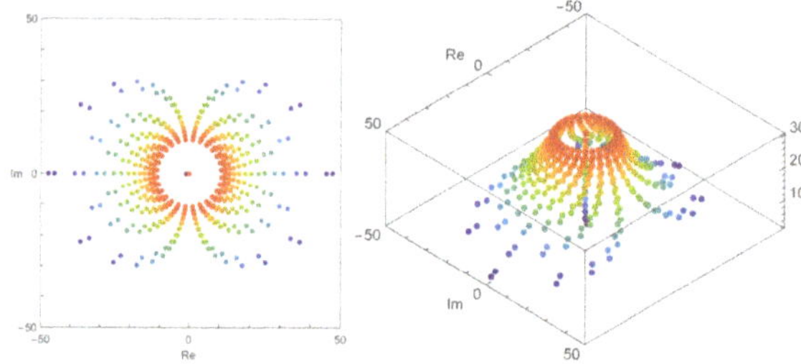

Figure 4. Stacking structure of approximate roots in the (p,q)-sine Bernoulli polynomials when $p = 0.1, q = 0.9$, and $y = 5$ in 3D.

5. Conclusions and Future Directions

In this paper, we explained about the (p,q)-cosine and (p,q)-sine Bernoulli polynomials, their basic properties, and various symmetric properties. Based on the above contents, we identified the structures of the approximate roots of the (p,q)-cosine and (p,q)-sine Bernoulli polynomials. As a result, we observed that the above polynomials obtain a structure of approximate roots, which has certain patterns and has a symmetric property under the given circumstances.

Further study is needed regarding whether the structure of approximate roots for the (p,q)-cosine and (p,q)-sine Bernoulli polynomials have symmetric properties under different circumstances. Furthermore, we think researching theories related to this topic is important to mathematicians.

Author Contributions: Conceptualization, J.Y.K.; Methodology, C.S.R.; Writing—original draft, J.Y.K. These autors contributed equally to this work. All authors have read and agreed to the published version of the manuscript.

Funding: This work was supported by the Basic Science Research Program through the National Research Foundation of Korea (NRF) funded by the Ministry of Science, ICT and Future Planning (No. 2017R1E1A1A03070483).

Acknowledgments: The authors would like to thank the referees for their valuable comments, which improved the original manuscript in its present form.

Conflicts of Interest: The authors declare that there is no conflict of interests regarding the publication of this paper.

References

1. Brodimas, G.; Jannussis, A.; Mignani, R. *Two-Parameter Quantum Groups*; Universita di Roma: Piazzale Aldo Moro, Roma, Italy, 1991; p. 820; View in KEK Scanned Document.
2. Chakrabarti, R.; Jagannathan, R. A (p,q)-oscillator realization of two-parameter quantum algebras. *J. Phys. A Math. Gen.* **1991**, *24*, L711–L718. [CrossRef]
3. Burban, I.M.; Klimyk, A.U. (p,q)-differentiation, (p,q)-integration and (p,q)-hypergeometric functions related to quantum groups. *Integral Transform. Spec. Funct.* **1994**, *2*, 15–36. [CrossRef]
4. Jagannathan, R. (P,Q)-special functions, special functions and differential equations. In *Proceedings of a Workshop*; The Institute of Mathematical Science: Matras, India, 1997; pp. 13–24.
5. Jagannathan, R.; Rao, K.S. Two-parameter quatum algebras, twin-basic numbers, and associated generalized hypergeometric series. In Proceedings of the Lnternational Conference on Number Theory and Mathematical Physics, Srinivasa Ramanujan Centre, Kumbakonam, India, 20–21 December 2005.
6. Corcino, R.B. On P,Q-Binomials coefficients. *Electron. J. Combin. Number Theory* **2008**, *8*, 1–16.
7. Sadjang, P.N. On the (p,q)-Gamma and the (p,q)-Beta functions. *arXiv* **2015**, arXiv:1506.07394v1.
8. Kupershmidt, B.A. Reflection Symmetries of q-Bernoulli Polynomials. *J. Nonlinear Math. Phys.* **2005**, *12*, 412–422. [CrossRef]
9. Kim, T.; Ryoo, C.S. Some identities for Euler and Bernoulli polynomials and their zeros. *Axioms* **2018**, *7*, 56. [CrossRef]
10. Kim, T. q-generalized Euler numbers and polynomials. *Russ. J. Math. Phys.* **2006**, *13*, 293–298. [CrossRef]
11. Kang, J.Y.; Ryoo, C.S. Various structures of the roots and explicit properties of q-cosine Bernoulli Polynomials and q-sine Bernoulli Polynomials. *Mathematics* **2020**, *8*, 463. [CrossRef]
12. Ryoo, C.S. A note on the zeros of the q-Bernoulli polynomials. *J. Appl. Math. Inform.* **2010**, *28*, 805–811.
13. Duran, U.; Acikgoz, M.; Araci, S. On some polynomials derived from (p,q)-calculus. *J. Comput. Theor. Nanosci.* **2016**, *13*, 7903–7908. [CrossRef]
14. Duran, U.; Acikgoz, M.; Araci, S. On (p,q)-Bernoulli, (p,q)-Euler and (p,q)-Genocchi polynomials. *J. Comput. Theor. Nanosci.* **2016**, *13*, 7833–7846. [CrossRef]
15. Acar, T.; Aral, A.; Mohiuddine, S.A. On Kantorovich modifivation of (p,q)-Baskakov operators. *J. Inequal. Appl.* **2016**, *98*, 1–14.
16. Arik, M.; Demircan, E.; Turgut, T.; Ekinci, L.; Mungan, M. Fibonacci Oscillators. *Z. Phys. C-Part. Fiels* **1992**, *55*, 89–95. [CrossRef]
17. Duran, U.; Acikgoz, M.; Araci, S. A study on some new results arising from (p,q)-calculus. *TWMS J. Pure Appl. Math.* **2020**, *11*, 57–71.
18. Jackson, H.F. On q-definite integrals. *Q. J. Pure Appl. Math.* **1910**, *41*, 193–203.
19. lSadjang, P.N. On the fundamental theorem of (p,q)-calculus and some (p,q)-Taylor formulas. *arXiv* **2013**, arXiv:1309.3934.
20. Wachs, M.; White, D. (p,q)-Stirling numbers and set partition statistics. *J. Combin. Theorey Ser. A* **1991**, *56*, 27–46. [CrossRef]

© 2020 by the authors. Licensee MDPI, Basel, Switzerland. This article is an open access article distributed under the terms and conditions of the Creative Commons Attribution (CC BY) license (http://creativecommons.org/licenses/by/4.0/).

Article

Explicit Properties of q-Cosine and q-Sine Euler Polynomials Containing Symmetric Structures

Cheon Seoung Ryoo [1] and Jung Yoog Kang [2],*

[1] Department of Mathematics, Hanman University, Daejeon 10216, Korea; ryoocs@hnu.kr
[2] Department of Mathematics Education, Silla University, Busan 469470, Korea
* Correspondence: jykang@silla.ac.kr

Received: 27 June 2020; Accepted: 25 July 2020; Published: 28 July 2020

Abstract: In this paper, we introduce q-cosine and q-sine Euler polynomials and determine identities for these polynomials. From these polynomials, we obtain some special properties using a power series of q-trigonometric functions, properties of q-exponential functions, and q-analogues of the binomial theorem. We investigate the approximate roots of q-cosine Euler polynomials that help us understand these polynomials. Moreover, we display the approximate roots movements of q-cosine Euler polynomials in a complex plane using the Newton method.

Keywords: q-cosine Euler polynomials; q-sine Euler polynomials; q-trigonometric function; q-exponential function

MSC: 11B68; 11B75; 33A10

---

1. Introduction

In 1990, Jackson who published influential papers on the subject introduced the q-number and its notation stems, see [1]. Floreanini and Vinet found that some properties of q-orthogonal polynomials are connected to the q-oscillator algebra in [1–4]. We begin by introducing several definitions related to q-numbers used in this paper, see [3,5–8].

Throughout this paper, the symbols, $\mathbb{N}, \mathbb{Z}, \mathbb{R}$ and $\mathbb{C}$ denotes the set of natural numbers, the set of integers, the set of real numbers and the set of complex numbers, respectively.

For $a \in \mathbb{C}$, $n \in \mathbb{N}$ and $|q| < 1$, the q-shifted factorial is defined by

$$(a;q)_0 = 1, \quad (a;q)_n = \prod_{j=0}^{n-1}(1 - q^j a), \quad (a;q)_\infty = \prod_{j=0}^{\infty}(1 - q^j a). \tag{1}$$

It is well known that

$$(a;q)_n = \sum_{k=0}^{n} \begin{bmatrix} n \\ k \end{bmatrix}_q q^{(1/2)k(k-1)}(-1)^k a^k. \tag{2}$$

Let $x, q \in \mathbb{R}$ with $q \neq 1$. The number

$$[x]_q = \frac{1 - q^x}{1 - q} \tag{3}$$

is called q-number. We note that $\lim_{q \to 1}[x]_q = x$. In particular, for $k \in \mathbb{Z}$, $[k]_q$ is a q-integer.

After the appearance of q-numbers, many mathematicians have studied topics such as q-differential equations, q-series, and q-trigonometric functions. Of course, mathematicians also constructed and researched q-Gaussian binomial coefficients, see [2–4,7–11].

Definition 1. *For $r \leq m$ and $m, r \in \mathbb{N}$, the q-Gaussian binomial coefficients are defined by*

$$\begin{bmatrix} m \\ r \end{bmatrix}_q = \frac{(q;q)_m}{(q;q)_{m-r}(q;q)_r}. \tag{4}$$

For $r = 0$, we note that $\begin{bmatrix} m \\ 0 \end{bmatrix}_q = 0$ with $0 \leq r \leq m$, and also, we note that $[n]_q! = [n]_q \cdots [2]_q[1]_q$ and $[0]_q! = 1$.

Definition 2. *Let $0 < |q| < 1$ and $|z| < \frac{1}{|1-q|}$. Then, the q-exponential function is defined by*

$$e_q(z) = \sum_{n=0}^{\infty} \frac{z^n}{[n]_q!} = \prod_{k=0}^{\infty} \frac{1}{(1 - (1-q)q^k z)}. \tag{5}$$

For $0 < q < 1$ and $|z| < \frac{2}{1-q}$, the other form of q-exponential function can be defined as

$$E_q(z) = e_{q^{-1}}(z) = \sum_{n=0}^{\infty} q^{\binom{n}{2}} \frac{z^n}{[n]_q!} = \prod_{k=0}^{\infty}(1 + (1-q)q^k z). \tag{6}$$

We note that $\lim_{q \to 1} e_q(z) = e^z$. Exponential function is expanded to the power series expressions of the two q-exponential functions by combining with q-numbers. Also, q-derivatives and q-integrals were extensively studied by many mathematicians, see [1,5,12]. Following the determination of the limit formulas for q-exponential functions taken from Rawlings [10], several other interesting q-series expansions were presented in the classical book by Andrews [5].

Theorem 1. *From Definition 2, we note that*

$$\begin{aligned} &(i) \quad e_q(x)e_q(y) = e_q(x+y), \quad if \quad yx = qxy. \\ &(ii) \quad e_q(x)E_q(-x) = 1. \end{aligned} \tag{7}$$

The proof of Theorem 1 and more properties of q-exponential functions can be found in [2,13].

Definition 3. *For real variable function f where $x \neq 0$, the q-derivative operator is defined as*

$$D_q f(x) = \frac{f(x) - f(qx)}{(1-q)x}. \tag{8}$$

We note that $D_q f(0) = f'(0)$. It is possible to prove that f is differentiable at 0 and it is clear that $D_q x^n = [n]_q x^{n-1}$.

In 2002, Kac and Pokman published a book about quantum calculus including q-derivatives and q-analogue of $(x-a)^n$ and q-trigonometric functions, see [14].

Definition 4. *Let n be a nonnegative integer. The q-analogues of subtraction and addition are defined by*

$$
\begin{aligned}
(i)\ (x \ominus a)_q^n &= \begin{cases} 1 & \text{if } n = 0 \\ (x - a)(x - qa) \cdots (x - q^{n-1}a) & \text{if } n \geq 1 \end{cases} \\
&= \sum_{k=0}^{n} \begin{bmatrix} n \\ k \end{bmatrix}_q \frac{(-1;q)_k(-1;q)_{n-k}}{2^n} x^k a^{n-k}, \\
(ii)\ (x \oplus a)_q^n &= \begin{cases} 1 & \text{if } n = 0 \\ (x + a)(x + qa) \cdots (x + q^{n-1}a) & \text{if } n \geq 1 \end{cases} \\
&= \sum_{k=0}^{n} \begin{bmatrix} n \\ k \end{bmatrix}_q \frac{(-1;q)_k(-1;q)_{n-k}}{2^n} x^k (-a)^{n-k},
\end{aligned}
\tag{9}
$$

respectively.

Definition 5. *Let $x \in \mathbb{R}$ and $i = \sqrt{-1}$. Then, the q-trigonometric functions are defined by*

$$
\begin{aligned}
sin_q(x) &= \frac{e_q(ix) - e_q(-ix)}{2i}, & SIN_q(x) &= \frac{E_q(ix) - E_q(-ix)}{2i} \\
cos_q(x) &= \frac{e_q(ix) + e_q(-ix)}{2}, & COS_q(x) &= \frac{E_q(ix) + E_q(-ix)}{2},
\end{aligned}
\tag{10}
$$

where $SIN_q(x) = sin_{q^{-1}}(x), COS_q(x) = cos_{q^{-1}}(x)$.

Theorem 2. *Using Theorem 1 (ii) and applying the chain rule, we have*

$$
(i)\quad COS_q(x)cos_q(x) + SIN_q(x)sin_q(x) = 1.
$$

$$
(ii)\quad \begin{cases} D_q sin_q(x) = cos_q(x), & D_q SIN_q(x) = COS_q(x), \\[2mm] D_q cos_q(x) = -sin_q(x), & D_q COS_q(x) = -SIN_q(x). \end{cases}
\tag{11}
$$

Moreover, we note Theorem 2 (i) is the q-analogue of the identity $sin^2 x + cos^2 x = 1$, see [3,4,7].

In 2004, Gasper and Rahman introduced a comprehensive account of the basic q-hypergeometric series, see [4]. During the last three decades, one of the bridges between science and applied mathematics has been q-calculus, see [10]. Based on the above concepts, many mathematicians explored various fields of mathematics including q-differential equations, q-series, q-hypergeometric functions, and q-gamma and q-beta functions. Moreover, various discrete distributions combining q-numbers can be found in [2]. Therefore, q-calculus plays an important role in many different areas of mathematics.

Many researchers who studied the Bernoulli, Euler, and Genocchi polynomials in various fields realized the important role of q-calculus in mathematics. For a long time, the topics of Bernoulli, Euler, and Genocchi polynomials have been extensively researched in many mathematical applications including analytical number theory, combinatorial analysis, p-adic analytic number theory, and other fields. Therefore, many mathematicians have started researching Bernoulli, Euler, and Genocchi polynomials combining q-numbers, see [9,13,15–21].

The following diagram briefly explains the relation between the various types of degenerate Euler polynomials, Euler polynomials, and q-Euler polynomials. The polynomials in the first row are researched by Calitz [21], Kim and Ryoo [16], respectively. The study of the second row of the diagram has brought about beneficial results in combinatorics and number theory. In particular, the cosine and sine Euler polynomials in the second row of the diagram contain a motive in this paper. This is because we hold several questions regarding what is a definition form of q-cosine and q-sine Euler polynomials and what is different properties between the q-cosine, q-sine Euler polynomials and the cosine, sine Euler polynomials.

The main subject of this paper is to construct q-cosine and q-sine Euler polynomials using Definitions 4 and 5. Also, we derive identities and properties for these polynomials in the third row of the diagram.

$$\frac{2}{(1+\lambda t)^{\frac{1}{\lambda}+1}}(1+\lambda t)^{\frac{x}{\lambda}} = \sum_{n=0}^{\infty} E_{n,\lambda}(x)\frac{t^n}{n!}$$
(degenerate Euler polynomials)

$$\frac{2}{(1+\lambda t)^{\frac{1}{\lambda}+1}}(1+\lambda t)^{\frac{x}{\lambda}}\cos(ty) = \sum_{n=0}^{\infty} E_{n,\lambda}^{(C)}(x,y)\frac{t^n}{n!}$$
(degenerate cosine Euler polynomials)

$$\frac{2}{(1+\lambda t)^{\frac{1}{\lambda}+1}}(1+\lambda t)^{\frac{x}{\lambda}}\sin(ty) = \sum_{n=0}^{\infty} E_{n,\lambda}^{(S)}(x,y)\frac{t^n}{n!}.$$
(degenerate sine Euler polynomials)

$$\frac{2}{e^t+1}e^{xt} = \sum_{n=0}^{\infty} E_n(x)\frac{t^n}{n!}$$
(Euler polynomials)

$$\frac{2}{e^t+1}e^{xt}\cos(ty) = \sum_{n=0}^{\infty} E_n^{(C)}(x,y)\frac{t^n}{n!}$$
(cosine Euler polynomials)

$$\frac{2}{e^t+1}e^{xt}\sin(ty) = \sum_{n=0}^{\infty} E_n^{(S)}(x,y)\frac{t^n}{n!}$$
(sine Euler polynomials)

$$\frac{2}{e_q(t)+1}e_q(xt) = \sum_{n=0}^{\infty} E_{n,q}(x)\frac{t^n}{n!}$$
(q-Euler polynomials)

$$\frac{2}{e_q(t)+1}e_q(xt)COS_q(ty) = \sum_{n=0}^{\infty} {}_C E_{n,q}(x,y)\frac{t^n}{n!}$$
(q-cosine Euler polynomials)

$$\frac{2}{e_q(t)+1}e_q(xt)SIN_q(ty) = \sum_{n=0}^{\infty} {}_S E_{n,q}(x,y)\frac{t^n}{n!}$$
(q-sine Euler polynomials)

The definition of q-Euler polynomials of the third row are as follows.

Definition 6. *The q-Euler numbers and polynomials are defined respectively as (see [19])*

$$\sum_{n=0}^{\infty} E_{n,q}\frac{t^n}{n!} = \frac{2}{e_q(t)+1}, \quad \sum_{n=0}^{\infty} E_{n,q}(z)\frac{t^n}{n!} = \frac{2}{e_q(t)+1}e_q(tz). \tag{12}$$

Recently, Kim and Ryoo introduced the basic concepts of cosine and sine Euler polynomials. In [16], the definitions and representative properties of cosine and sine Euler polynomials are as follows.

Definition 7. *The cosine Euler polynomials $E_n^{(C)}(x,y)$ and the sine Euler polynomials $E_n^{(S)}(x,y)$ are defined by means of the generating functions*

$$\sum_{n=0}^{\infty} E_n^{(C)}(x,y)\frac{t^n}{n!} = \frac{2}{e^t+1}e^{tx}\cos(ty), \quad \sum_{n=0}^{\infty} E_n^{(S)}(x,y)\frac{t^n}{n!} = \frac{2}{e^t+1}e^{tx}\sin(ty). \tag{13}$$

In this paper, we denote that $E_n^{(C)}(x,y) = {}_C\mathcal{E}_n(x,y)$ and $E_n^{(S)}(x,y) = {}_S\mathcal{E}_n(x,y)$.

Theorem 3. *For $n \geq 1$, we have*

$$
\begin{aligned}
&\text{(i)} \quad E_n^{(C)}(x+1,y) - E_n^{(C)}(x,y) = 2C_n(x,y), \\
&\text{(ii)} \quad E_n^{(S)}(x+1,y) - E_n^{(S)}(x,y) = 2S_n(x,y).
\end{aligned}
\tag{14}
$$

Based on [16], which contains Definition 7 and Theorem 3, many researchers found various expanded numbers and polynomials and their identities, see [15,20].

The main goal of this paper is to find various properties of q-cosine and q-sine Euler polynomials such as addition theorem, partial q-derivative, basic symmetric properties so on. In Section 2, we construct q-cosine and q-sine Euler polynomials. Then, using q-calculus, we identify basic properties of these polynomials. Section 3 presents an investigation of the special properties of q-cosine and q-sine Euler polynomials such as the identity of q-sine Euler polynomials using q-analogues of subtraction and addition. This is based on the properties of q-trigonometric and q-exponential functions. Moreover, we derive relationships between q-cosine and q-sine Euler polynomials and q-cosine and q-sine Bernoulli polynomials. In Section 4, we display the structure of approximate roots for q-cosine and q-sine Euler polynomials and find properties of these polynomials. We present some figures of the approximate roots of these polynomials in a complex plane using Newton's method.

2. Some Basic Properties of q-cosine and q-sine Euler Polynomials

In this section, we construct the q-cosine and q-sine Euler polynomials by using Theorem 4. From the generating functions of these polynomials, we obtain some basic properties and identities. Moreover, we derive symmetric properties and partial q-derivatives for q-cosine and q-sine Euler polynomials.

Definition 8. *The generating functions of q-cosine Euler polynomials and q-sine Euler polynomials are correspondingly defined by*

$$
\sum_{n=0}^{\infty} {}_C\mathcal{E}_{n,q}(x,y) \frac{t^n}{[n]_q!} = \frac{2}{e_q(t)+1} e_q(tx) COS_q(ty)
$$

and $\qquad\qquad\qquad\qquad\qquad\qquad\qquad\qquad\qquad\qquad\qquad\qquad$ (15)

$$
\sum_{n=0}^{\infty} {}_S\mathcal{E}_{n,q}(x,y) \frac{t^n}{[n]_q!} = \frac{2}{e_q(t)+1} e_q(tx) SIN_q(ty).
$$

From Definition 8, q-sine Euler polynomials can be confirmed as the following:

$$
\begin{aligned}
{}_S\mathcal{E}_{0,q}(x,y) &= 0, \\
{}_S\mathcal{E}_{1,q}(x,y) &= -\frac{y}{1+q}, \\
{}_S\mathcal{E}_{2,q}(x,y) &= -\frac{(1+q+q^2-2x)y}{2(1+q+q^2)}, \\
{}_S\mathcal{E}_{3,q}(x,y) &= -\frac{y(1+2x+2qx-4x^2+q^2(1+2x)+q^5(-1+4y^2)+q^3(-1+2x+4y^2))}{4(1+q)(1+q^2)}, \\
&\cdots.
\end{aligned}
$$

We will introduce the certain form of q-cosine Euler polynomials in Section 4. The motivation to derive the definition of q-cosine Euler polynomials and q-sine Euler polynomials can be found in Theorem 4.

Theorem 4. *For $x, y \in \mathbb{R}$ and $i = \sqrt{-1}$, we have*

$$(i) \quad \sum_{n=0}^{\infty} \left(\frac{\mathcal{E}_{n,q}((x \oplus iy)_q) + \mathcal{E}_{n,q}((x \ominus iy)_q)}{2} \right) \frac{t^n}{[n]_q!} = \frac{2}{e_q(t)+1} e_q(tx) COS_q(ty),$$

$$(ii) \quad \sum_{n=0}^{\infty} \left(\frac{\mathcal{E}_{n,q}((x \oplus iy)_q) - \mathcal{E}_{n,q}((x \ominus iy)_q)}{2i} \right) \frac{t^n}{[n]_q!} = \frac{2}{e_q(t)+1} e_q(tx) SIN_q(ty).$$

$$(16)$$

Proof. (i) We defined the generating function of q-Euler polynomials in Definition 6. Let us substitute $(x \oplus iy)_q$ instead of z in the q-Euler polynomials. By using a property of q-analogues of the sine and cosine functions and by using Definition 1, we can find

$$
\begin{aligned}
\sum_{n=0}^{\infty} \mathcal{E}_{n,q}((x \oplus iy)_q) \frac{t^n}{[n]_q!} &= \frac{2}{e_q(t)+1} \sum_{n=0}^{\infty} (x \oplus iy)_q^n \frac{t^n}{[n]_q!} \\
&= \frac{2}{e_q(t)+1} \sum_{n=0}^{\infty} \left(\sum_{k=0}^{n} \begin{bmatrix} n \\ k \end{bmatrix}_q q^{\binom{n-k}{2}} x^k (iy)^{n-k} \right) \frac{t^n}{[n]_q!} \\
&= \frac{2}{e_q(t)+1} e_q(tx) E_q(ity) \\
&= \frac{2}{e_q(t)+1} e_q(tx) \left(COS_q(ty) + iSIN_q(ty) \right).
\end{aligned}
$$

$$(17)$$

By using a similar method as when finding Equation (17), we can also obtain

$$
\begin{aligned}
\sum_{n=0}^{\infty} \mathcal{E}_{n,q}((x \ominus iy)_q) \frac{t^n}{[n]_q!} &= \frac{2}{e_q(t)+1} e_q(tx) E_q(-ity) \\
&= \frac{2}{e_q(t)+1} e_q(tx) \left(COS_q(ty) - iSIN_q(ty) \right).
\end{aligned}
$$

$$(18)$$

(ii) We can prove Theorem 4 (ii) through Equations (17) and (18). $\square$

Remark 1. *From the Theorem 4 and Definition 8, the following holds*

$$(i) \quad {}_C\mathcal{E}_{n,q}(x,y) = \frac{\mathcal{E}_{n,q}((x \oplus iy)_q) + \mathcal{E}_{n,q}((x \ominus iy)_q)}{2}$$

$$(ii) \quad {}_S\mathcal{E}_{n,q}(x,y) = \frac{\mathcal{E}_{n,q}((x \oplus iy)_q) - \mathcal{E}_{n,q}((x \ominus iy)_q)}{2i}.$$

$$(19)$$

In [15], $C_{n,q}(x,y)$ and $S_{n,q}(x,y)$ are defined as follows:

$$(i) \quad \sum_{n=0}^{\infty} C_{n,q}(x,y) \frac{t^n}{[n]_q!} = e_q(tx) COS_q(ty),$$

$$(ii) \quad \sum_{n=0}^{\infty} S_{n,q}(x,y) \frac{t^n}{[n]_q!} = e_q(tx) SIN_q(ty).$$

$$(20)$$

Theorem 5. *For any real numbers x, y, we have*

$$(i) \quad {}_C\mathcal{E}_{n,q}(x,y) = \sum_{k=0}^{n} \begin{bmatrix} n \\ k \end{bmatrix}_q E_{k,q} C_{n-k,q}(x,y),$$

$$(ii) \quad {}_S\mathcal{E}_{n,q}(x,y) = \sum_{k=0}^{n} \begin{bmatrix} n \\ k \end{bmatrix}_q E_{k,q} S_{n-k,q}(x,y),$$

$$(21)$$

where $E_{n,q}$ is the q-Euler numbers.

Proof. (i) From the generating function of q-cosine Euler polynomials, we can find a relation between the q-Euler numbers and $C_{n,q}(x,y)$ as follows.

$$
\begin{aligned}
\sum_{n=0}^{\infty} {}_{c}\mathcal{E}_{n,q}(x,y)\frac{t^n}{[n]_q!} &= \sum_{n=0}^{\infty} {}_{c}E_{n,q}\frac{t^n}{[n]_q!} \sum_{n=0}^{\infty} C_{n,q}(x,y)\frac{t^n}{[n]_q!} \\
&= \sum_{n=0}^{\infty} \left(\sum_{k=0}^{n} \begin{bmatrix} n \\ k \end{bmatrix}_q E_{k,q} C_{n-k,q}(x,y) \right) \frac{t^n}{[n]_q!},
\end{aligned}
\tag{22}
$$

and we obtain the required result of Theorem 5 (i).

(ii) We also find a relation between the q-Euler numbers and $S_{n,q}(x,y)$ in a similar way as in the proof of (i) and we have the required result. $\square$

Theorem 6. *For a nonnegative integer k and $|q| < 1$, we derive*

$$
\begin{aligned}
(i) \quad {}_{c}\mathcal{E}_{n,q}(x,y) &= \sum_{k=0}^{\left[\frac{n}{2}\right]} \begin{bmatrix} n \\ 2k \end{bmatrix}_q (-1)^k q^{(2k-1)k} y^{2k} E_{n-2k,q}(x), \\
(ii) \quad {}_{s}\mathcal{E}_{n,q}(x,y) &= \sum_{k=0}^{\left[\frac{n-1}{2}\right]} \begin{bmatrix} n \\ 2k+1 \end{bmatrix}_q (-1)^k q^{(2k+1)k} y^{2k+1} E_{n-(2k+1),q}(x),
\end{aligned}
\tag{23}
$$

where $[x]$ is the greatest integer not exceeding x and $E_{n,q}(x)$ is the q-Euler polynomials.

Proof. (i) From the generating function of q-Euler polynomials, we can change the q-cosine Euler polynomials as follows

$$
\sum_{n=0}^{\infty} {}_{c}\mathcal{E}_{n,q}(x,y)\frac{t^n}{[n]_q!} = \sum_{n=0}^{\infty} E_{n,q}(x)\frac{t^n}{[n]_q!} COS_q(ty).
\tag{24}
$$

By using the power series of $COS_q(x)$, the right-hand side of Equation (24) is transformed as

$$
\begin{aligned}
&\sum_{n=0}^{\infty} E_{n,q}(x)\frac{t^n}{[n]_q!} \sum_{n=0}^{\infty} (-1)^n q^{(2n-1)n} y^{2n} \frac{t^{2n}}{[2n]_q!} \\
&= \sum_{n=0}^{\infty} \left(\sum_{k=0}^{n} \begin{bmatrix} n+k \\ 2k \end{bmatrix}_q (-1)^k q^{(2k-1)k} y^{2k} E_{n-k,q}(x) \right) \frac{t^{n+k}}{[n+k]_q!} \\
&= \sum_{n=0}^{\infty} \left(\sum_{k=0}^{\left[\frac{n}{2}\right]} \begin{bmatrix} n \\ 2k \end{bmatrix}_q (-1)^k q^{(2k-1)k} y^{2k} E_{n-2k,q}(x) \right) \frac{t^n}{[n]_q!},
\end{aligned}
\tag{25}
$$

and we complete the proof of Theorem 6 (i).

(ii) By applying the power series of $SIN_q(x)$ in the generating function of q-sine Euler polynomials, we have

$$
\begin{aligned}
\sum_{n=0}^{\infty} {}_{s}\mathcal{E}_{n,q}(x,y)\frac{t^n}{[n]_q!} &= \sum_{n=0}^{\infty} E_{n,q}(x)\frac{t^n}{[n]_q!} \sum_{n=0}^{\infty} (-1)^n q^{(2n+1)n} y^{2n+1} \frac{t^{2n+1}}{[2n+1]_q!} \\
&= \sum_{n=0}^{\infty} \left(\sum_{k=0}^{\left[\frac{n-1}{2}\right]} \begin{bmatrix} n \\ 2k+1 \end{bmatrix}_q (-1)^k q^{(2k+1)k} y^{2k+1} E_{n-2(k+1),q}(x) \right) \frac{t^n}{[n]_q!},
\end{aligned}
\tag{26}
$$

and we finish the proof of Theorem 6 (ii). $\square$

Corollary 1. *Let $y = 1$ in Theorem 6. Then, the following holds*

$$
(i) \quad {}_C\mathcal{E}_{n,q}(x,1) = \sum_{k=0}^{[\frac{n}{2}]} \begin{bmatrix} n \\ 2k \end{bmatrix}_q (-1)^k q^{(2k-1)k} E_{n-2k,q}(x),
$$

$$
(ii) \quad {}_S\mathcal{E}_{n,q}(x,1) = \sum_{k=0}^{[\frac{n-1}{2}]} \begin{bmatrix} n \\ 2k+1 \end{bmatrix}_q (-1)^k q^{(2k+1)k} E_{n-(2k+1),q}(x),
$$

$$\tag{27}$$

where $[x]$ is the greatest integer not exceeding x and $E_{n,q}(x)$ is the q-Euler polynomials.

Theorem 7. *Let $x,y \in \mathbb{R}$, $|q| < 1$, and $e_q(t) \neq -1$. Then, we have*

$$
(i) \quad C_{n,q}(x,y) = \frac{1}{2} \left(\sum_{k=0}^{n} \begin{bmatrix} n \\ k \end{bmatrix}_q {}_C\mathcal{E}_{k,q}(x,y) + {}_C\mathcal{E}_{n,q}(x,y) \right),
$$

$$
(ii) \quad S_{n,q}(x,y) = \frac{1}{2} \left(\sum_{k=0}^{n} \begin{bmatrix} n \\ k \end{bmatrix}_q {}_S\mathcal{E}_{k,q}(x,y) + {}_S\mathcal{E}_{n,q}(x,y) \right).
$$

$$\tag{28}$$

Proof. (i) When $e_q(t) \neq -1$, we can consider the generating function of the q-cosine Euler polynomials to be

$$
\sum_{n=0}^{\infty} {}_C\mathcal{E}_{n,q}(x,y) \frac{t^n}{[n]_q!} (e_q(t) + 1) = 2e_q(tx)COS_q(ty).
$$

$$\tag{29}$$

The left-hand side of Equation (29) is changed as follows.

$$
\sum_{n=0}^{\infty} \left(\sum_{k=0}^{n} \begin{bmatrix} n \\ k \end{bmatrix}_q {}_C\mathcal{E}_{k,q}(x,y) + {}_C\mathcal{E}_{n,q}(x,y) \right) \frac{t^n}{[n]_q!} = \sum_{n=0}^{\infty} {}_C\mathcal{E}_{n,q}(x,y) \frac{t^n}{[n]_q!} (e_q(t) + 1).
$$

$$\tag{30}$$

The right-hand side of (29) is transformed as

$$
2e_q(tx)COS_q(ty) = 2 \sum_{n=0}^{\infty} C_{n,q}(x,y) \frac{t^n}{[n]_q!}.
$$

$$\tag{31}$$

By using Equations (30) and (31), we find the required result.
(ii) In a similar method as in the proof of (i), we have

$$
\sum_{n=0}^{\infty} {}_S\mathcal{E}_{n,q}(x,y) \frac{t^n}{[n]_q!} (e_q(t) + 1) = 2e_q(tx)SIN_q(ty).
$$

$$\tag{32}$$

Then, we obtain

$$
\sum_{n=0}^{\infty} \left(\sum_{k=0}^{n} \begin{bmatrix} n \\ k \end{bmatrix}_q {}_S\mathcal{E}_{k,q}(x,y) + {}_S\mathcal{E}_{n,q}(x,y) \right) \frac{t^n}{[n]_q!} = 2 \sum_{n=0}^{\infty} S_{n,q}(x,y) \frac{t^n}{[n]_q!}.
$$

$$\tag{33}$$

Therefore, we finish the proof of Theorem 7 (ii). $\square$

Corollary 2. *If $q \to 1$ in Theorem 7, we have*

$$
\begin{aligned}
(i) \quad & C_n(x,y) = \frac{1}{2}\left(\sum_{k=0}^{n} \binom{n}{k} {}_c\mathcal{E}_k(x,y) + {}_c\mathcal{E}_n(x,y) \right), \\
(ii) \quad & S_n(x,y) = \frac{1}{2}\left(\sum_{k=0}^{n} \binom{n}{k} {}_s\mathcal{E}_k(x,y) + {}_s\mathcal{E}_n(x,y) \right),
\end{aligned}
\tag{34}
$$

where ${}_c\mathcal{E}_n(x,y)$ is the cosine Euler polynomials and ${}_s\mathcal{E}_n(x,y)$ is the sine Euler polynomials.

Theorem 8. *For any real number x, y and $|q| < 1$, we have*

$$
\begin{aligned}
(i) \quad & \frac{\partial}{\partial x}{}_c\mathcal{E}_{n,q}(x,y) = [n]_q {}_c\mathcal{E}_{n-1,q}(x,y), \quad \frac{\partial}{\partial y}{}_c\mathcal{E}_{n,q}(x,y) = -[n]_q {}_s\mathcal{E}_{n-1,q}(x,qy). \\
(ii) \quad & \frac{\partial}{\partial x}{}_s\mathcal{E}_{n,q}(x,y) = [n]_q {}_s\mathcal{E}_{n-1,q}(x,y), \quad \frac{\partial}{\partial y}{}_s\mathcal{E}_{n,q}(x,y) = [n]_q {}_c\mathcal{E}_{n-1,q}(x,qy).
\end{aligned}
\tag{35}
$$

Proof. (i) For any real number x, we can find the partial q-derivative for q-cosine Euler polynomials as

$$
\sum_{n=0}^{\infty} \frac{\partial}{\partial x}{}_c\mathcal{E}_{n,q}(x,y)\frac{t^n}{[n]_q!} = \frac{2}{e_q(t)+1}COS_q(ty)\frac{\partial}{\partial x}e_q(tx) = \frac{2t}{e_q(t)+1}e_q(tx)COS_q(ty).
\tag{36}
$$

Using the q-derivative of the q-cosine function, we find

$$
D_q COS_q(ty) = -tSIN_q(qty).
\tag{37}
$$

From Equation (37), we get

$$
\sum_{n=0}^{\infty} \frac{\partial}{\partial y}{}_c\mathcal{E}_{n,q}(x,y)\frac{t^n}{[n]_q!} = \frac{2}{e_q(t)+1}e_q(tx)\frac{\partial}{\partial y}COS_q(ty) = -\frac{2t}{e_q(t)+1}e_q(tx)SIN_q(qty)
\tag{38}
$$

and we obtain the required results.

(ii) We also consider the partial q-derivative of q-sine Euler polynomials as

$$
\sum_{n=0}^{\infty} \frac{\partial}{\partial x}{}_s\mathcal{E}_{n,q}(x,y)\frac{t^n}{[n]_q!} = \frac{2t}{e_q(t)+1}e_q(tx)SIN_q(ty),
\tag{39}
$$

and note that $D_q SIN_q(ty) = tCOS_q(qty)$. Then, we obtain the results of Theorem 8. $\quad\square$

In [18], Liu and Wang studied some symmetric properties of the Bernoulli and Euler polynomials. Based on the above paper, we observe some symmetric properties of the q-cosine and q-sine Euler polynomials. Moreover, symmetric properties can be found in the cosine and sine Euler polynomials.

Theorem 9. *For any integers a, b, we have*

$$
\begin{aligned}
(i) \quad & \sum_{k=0}^{n} \begin{bmatrix} n \\ k \end{bmatrix}_q a^{n-k} b^k {}_C\mathcal{E}_{n-k,q}(bx, by) {}_C\mathcal{E}_{k,q}(ax, ay) \\
= & \sum_{k=0}^{n} \begin{bmatrix} n \\ k \end{bmatrix}_q b^{n-k} a^k {}_C\mathcal{E}_{n-k,q}(ax, ay) {}_C\mathcal{E}_{k,q}(bx, by), \\
(ii) \quad & \sum_{k=0}^{n} \begin{bmatrix} n \\ k \end{bmatrix}_q a^{n-k} b^k {}_S\mathcal{E}_{n-k,q}(bx, by) {}_S\mathcal{E}_{k,q}(ax, ay) \\
= & \sum_{k=0}^{n} \begin{bmatrix} n \\ k \end{bmatrix}_q b^{n-k} a^k {}_S\mathcal{E}_{n-k,q}(ax, ay) {}_S\mathcal{E}_{k,q}(bx, by).
\end{aligned}
\tag{40}
$$

Proof. (i) To find a symmetric property, we assume form A such that

$$
A := \frac{4 \left(e_q(abtx) COS_q(abty)\right)^2}{(e_q(at) + 1)(e_q(bt) + 1)}.
\tag{41}
$$

By considering the generating function of q-cosine Euler polynomials in Equation (41), we can find the following equation:

$$
\begin{aligned}
A &= \frac{2}{e_q(at) + 1} e_q(abtx) COS_q(abty) \frac{2}{e_q(bt) + 1} e_q(abtx) COS_q(abty) \\
&= \sum_{n=0}^{\infty} \left(\sum_{k=0}^{n} \begin{bmatrix} n \\ k \end{bmatrix}_q a^{n-k} b^k {}_C\mathcal{E}_{n-k,q}(bx, by) {}_C\mathcal{E}_{k,q}(ax, ay) \right) \frac{t^n}{[n]_q!},
\end{aligned}
\tag{42}
$$

and

$$
\begin{aligned}
A &= \frac{2}{e_q(bt) + 1} e_q(abtx) COS_q(abty) \frac{2}{e_q(at) + 1} e_q(abtx) COS_q(abty) \\
&= \sum_{n=0}^{\infty} \left(\sum_{k=0}^{n} \begin{bmatrix} n \\ k \end{bmatrix}_q b^{n-k} a^k {}_C\mathcal{E}_{n-k,q}(ax, ay) {}_C\mathcal{E}_{k,q}(bx, by) \right) \frac{t^n}{[n]_q!}.
\end{aligned}
\tag{43}
$$

From Equations (42) and (43), we can find the required result.

(ii) In a similar way as with form A, we can make form A' such that

$$
A' := \frac{4 \left(e_q(abtx) SIN_q(abty)\right)^2}{(e_q(at) + 1)(e_q(bt) + 1)},
\tag{44}
$$

and we can find Theorem 9 (ii) in a same manner as (i). $\square$

Corollary 3. *Assume $a = 1$ in Theorem 9. Then, the following holds*

$$
\begin{aligned}
(i) \quad & \sum_{k=0}^{n} \begin{bmatrix} n \\ k \end{bmatrix}_q b^k {}_C\mathcal{E}_{n-k,q}(bx, by) {}_C\mathcal{E}_{k,q}(x, y) = \sum_{k=0}^{n} \begin{bmatrix} n \\ k \end{bmatrix}_q b^{n-k} {}_C\mathcal{E}_{n-k,q}(x, y) {}_C\mathcal{E}_{k,q}(bx, by), \\
(ii) \quad & \sum_{k=0}^{n} \begin{bmatrix} n \\ k \end{bmatrix}_q b^k {}_S\mathcal{E}_{n-k,q}(bx, by) {}_S\mathcal{E}_{k,q}(x, y) = \sum_{k=0}^{n} \begin{bmatrix} n \\ k \end{bmatrix}_q b^{n-k} {}_S\mathcal{E}_{n-k,q}(x, y) {}_S\mathcal{E}_{k,q}(bx, by).
\end{aligned}
\tag{45}
$$

Corollary 4. *Assume* $q \to 1$ *in Theorem 9. Then, the following holds*

$$(i) \sum_{k=0}^{n} \binom{n}{k} a^{n-k} b^{k} {}_C\mathcal{E}_{n-k}(bx, by) {}_C\mathcal{E}_{k}(ax, ay) = \sum_{k=0}^{n} \binom{n}{k} b^{n-k} a^{k} {}_C\mathcal{E}_{n-k}(ax, ay) {}_C\mathcal{E}_{k}(bx, by),$$

$$(ii) \sum_{k=0}^{n} \binom{n}{k} a^{n-k} b^{k} {}_S\mathcal{E}_{n-k}(bx, by) {}_S\mathcal{E}_{k}(ax, ay) = \sum_{k=0}^{n} \binom{n}{k} b^{n-k} a^{k} {}_S\mathcal{E}_{n-k}(ax, ay) {}_S\mathcal{E}_{k}(bx, by),$$

$$(46)$$

where ${}_C\mathcal{E}_n$ *is the cosine Euler polynomials and* ${}_S\mathcal{E}_n$ *is the sine Euler polynomials.*

Theorem 10. *For any integers* $a, b,$ *and* $|q| < 1.$ *Then, we obtain*

$$\sum_{k=0}^{n} \begin{bmatrix} n \\ k \end{bmatrix}_q a^{n-k} b^{k} {}_C\mathcal{E}_{n-k,q}(bx, by) {}_S\mathcal{E}_{k,q}(ax, ay)$$

$$= \sum_{k=0}^{n} \begin{bmatrix} n \\ k \end{bmatrix}_q b^{n-k} a^{k} {}_C\mathcal{E}_{n-k,q}(ax, ay) {}_S\mathcal{E}_{k,q}(bx, by).$$

$$(47)$$

Proof. To derive a symmetric relation mixing the q-cosine Euler polynomials and the q-sine Euler polynomials, we take form B as the following.

$$B := \frac{4 COS_q(abty) SIN_q(abty) (e_q(abtx))^2}{(e_q(at) + 1)(e_q(bt) + 1)}.$$

$$(48)$$

From form B, we can find the following equations:

$$B = \frac{2}{e_q(at) + 1} e_q(abtx) COS_q(abty) \frac{2}{e_q(bt) + 1} e_q(abtx) SIN_q(abty)$$

$$= \sum_{n=0}^{\infty} \left(\sum_{k=0}^{n} \begin{bmatrix} n \\ k \end{bmatrix}_q a^{n-k} b^{k} {}_C\mathcal{E}_{n-k,q}(bx, by) {}_S\mathcal{E}_{k,q}(ax, ay) \right) \frac{t^n}{[n]_q!},$$

$$(49)$$

and

$$B = \frac{2}{e_q(bt) + 1} e_q(abtx) COS_q(abty) \frac{2}{e_q(at) + 1} e_q(abtx) SIN_q(abty)$$

$$= \sum_{n=0}^{\infty} \left(\sum_{k=0}^{n} \begin{bmatrix} n \\ k \end{bmatrix}_q b^{n-k} a^{k} {}_C\mathcal{E}_{n-k,q}(ax, ay) {}_S\mathcal{E}_{k,q}(bx, by) \right) \frac{t^n}{[n]_q!}.$$

$$(50)$$

From (49) and (50), we can immediately complete the proof of Theorem 10. $\square$

Corollary 5. *Suppose* $q \to 1$ *in Theorem 10. Then the following holds*

$$\sum_{k=0}^{n} \binom{n}{k} a^{n-k} b^{k} {}_C\mathcal{E}_{n-k}(bx, by) {}_S\mathcal{E}_{k}(ax, ay) = \sum_{k=0}^{n} \binom{n}{k} b^{n-k} a^{k} {}_C\mathcal{E}_{n-k}(ax, ay) {}_S\mathcal{E}_{k}(bx, by). \quad (51)$$

3. Some Special Properties of q-cosine Euler Polynomials and q-sine Euler Polynomials

In this section, we obtain some special properties of q-cosine and q-sine Euler polynomials using the properties of q-trigonometric functions, $(x \oplus y)_q$, and so on. Moreover, we find various types of relationships between q-cosine, sine Euler polynomials and other polynomials.

Theorem 11. *For $|q| < 1$, we obtain*

$$(i) \quad {}_c\mathcal{E}_{n,q}(1,y) = \sum_{k=0}^{n} \begin{bmatrix} n \\ k \end{bmatrix}_q (-1)^{n-k} q^{\binom{n-k}{2}} \left(2C_{k,q}(x,y) - {}_c\mathcal{E}_{k,q}(x,y) \right) x^{n-k},$$

$$(ii) \quad {}_s\mathcal{E}_{n,q}(1,y) = \sum_{k=0}^{n} \begin{bmatrix} n \\ k \end{bmatrix}_q (-1)^{n-k} q^{\binom{n-k}{2}} (2S_{k,q}(x,y) - {}_s\mathcal{E}_{k,q}(x,y)) x^{n-k}. \tag{52}$$

Proof. (i) When $x = 1$ in the generating function of q-cosine Euler polynomials, ${}_c\mathcal{E}_{n,q}(x,y)$, we have

$$\sum_{n=0}^{\infty} {}_c\mathcal{E}_{n,q}(1,y) \frac{t^n}{[n]_q!} = 2COS_q(ty) - \frac{2}{e_q(t)+1} COS_q(ty). \tag{53}$$

Using $e_q(x)E_q(-x) = 1$ and $[n]_{q^{-1}}! = q^{-\binom{n}{2}}[n]_q!$, the left-hand side of Equation (53) can be written as the following:

$$\sum_{n=0}^{\infty} {}_c\mathcal{E}_{n,q}(1,y) \frac{t^n}{[n]_q!} = \left(2e_q(tx)COS_q(ty) - \frac{2}{e_q(t)+1} e_q(tx)COS_q(ty) \right) E_q(-tx)$$

$$= \sum_{n=0}^{\infty} (2C_{n,q}(x,y) - {}_c\mathcal{E}_{n,q}(x,y)) \frac{t^n}{[n]_q!} \sum_{n=0}^{\infty} (-1)^n q^{\binom{n}{2}} x^n \frac{t^n}{[n]_q!}$$

$$= \sum_{n=0}^{\infty} \left(\sum_{k=0}^{n} \begin{bmatrix} n \\ k \end{bmatrix}_q (-1)^{n-k} q^{\binom{n-k}{2}} (2C_{k,q}(x,y) - {}_c\mathcal{E}_{k,q}(x,y)) x^{n-k} \right) \frac{t^n}{[n]_q!}. \tag{54}$$

Comparing the both sides of Equation (54), we obtain the required result.

(ii) By using the same method as in the proof of (i), we have the proof of Theorem 11 (ii). $\square$

Corollary 6. *When $q \to 1$ in Theorem 11, the following holds*

$$(i) \quad {}_c\mathcal{E}_n(1,y) = \sum_{k=0}^{n} \binom{n}{k} (-1)^{n-k} \left((2C_k(x,y) - {}_c\mathcal{E}_k(x,y)) \right) x^{n-k},$$

$$(ii) \quad {}_s\mathcal{E}_n(1,y) = \sum_{k=0}^{n} \binom{n}{k} (-1)^{n-k} (2S_k(x,y) - {}_s\mathcal{E}_k(x,y)) x^{n-k}, \tag{55}$$

where ${}_c\mathcal{E}_n(x,y)$ is the cosine Euler polynomials and ${}_s\mathcal{E}_n(x,y)$ is the sine Euler polynomials.

Lemma 1. *For $|q| < 1$ and a real number r, we have*

$$(i) \quad {}_c\mathcal{E}_{n,q}((x \oplus r)_q, y) = \sum_{k=0}^{n} \begin{bmatrix} n \\ k \end{bmatrix}_q q^{\binom{n-k}{2}} {}_c\mathcal{E}_{k,q}(x,y) r^{n-k},$$

$$(ii) \quad {}_c\mathcal{E}_{n,q}((x \ominus r)_q, y) = \sum_{k=0}^{n} \begin{bmatrix} n \\ k \end{bmatrix}_q (-1)^{n-k} q^{\binom{n-k}{2}} {}_c\mathcal{E}_{k,q}(x,y) r^{n-k},$$

$$(iii) \quad {}_s\mathcal{E}_{n,q}((x \oplus r)_q, y) = \sum_{k=0}^{n} \begin{bmatrix} n \\ k \end{bmatrix}_q q^{\binom{n-k}{2}} {}_s\mathcal{E}_{k,q}(x,y) r^{n-k},$$

$$(iv) \quad {}_s\mathcal{E}_{n,q}((x \ominus r)_q, y) = \sum_{k=0}^{n} \begin{bmatrix} n \\ k \end{bmatrix}_q (-1)^{n-k} q^{\binom{n-k}{2}} {}_s\mathcal{E}_{k,q}(x,y) r^{n-k}. \tag{56}$$

Proof. (i) By substituting $(x \oplus r)_q$ instead of x in the generating function of q-cosine Euler polynomials and using q-exponential functions, we derive

$$\sum_{n=0}^{\infty} {}_c\mathcal{E}_{n,q}((x \oplus r)_q, y) \frac{t^n}{[n]_q!} = \frac{2}{e_q(t)+1} e_q(tx) COS_q(ty) E_q(tr)$$

$$= \sum_{n=0}^{\infty} \left(\sum_{k=0}^{n} \begin{bmatrix} n \\ k \end{bmatrix}_q q^{\binom{n-k}{2}} {}_c\mathcal{E}_{k,q}(x,y) r^{n-k} \right) \frac{t^n}{[n]_q!},$$

(57)

Thus, we find the required result immediately.

(ii) Putting $(x \ominus r)_q$ into x in the generating function of q-cosine Euler polynomials and using q-exponential functions, we have

$$\sum_{n=0}^{\infty} {}_c\mathcal{E}_{n,q}((x \ominus r)_q, y) \frac{t^n}{[n]_q!} = \frac{2}{e_q(t)+1} e_q(tx) COS_q(ty) E_q(-tr). \tag{58}$$

We find the required result in a similar way as in the proof of (i).

(iii) We consider that

$$\sum_{n=0}^{\infty} {}_s\mathcal{E}_{n,q}((x \oplus r)_q, y) \frac{t^n}{[n]_q!} = \frac{2}{e_q(t)+1} e_q(tx) SIN_q(ty) E_q(tr). \tag{59}$$

Then, we have the following result.

(iv) If we set $(x \ominus r)$ in x in the generating function of q-sine Euler polynomials, then we have

$$\sum_{n=0}^{\infty} {}_s\mathcal{E}_{n,q}((x \ominus r)_q, y) \frac{t^n}{[n]_q!} = \frac{2}{e_q(t)+1} e_q(tx) SIN_q(ty) E_q(-tr). \tag{60}$$

From Equation (60), we obtain the desired result. $\square$

Theorem 12. *Let* $|q| < 1$ *and* $r, x, y \in \mathbb{R}$. *From the Lemma 1, we have*

(i) $\quad {}_c\mathcal{E}_{n,q}((x \oplus r)_q, y) + {}_c\mathcal{E}_{n,q}((x \ominus r)_q, y)$

$$= \begin{cases} 2\sum_{k=0}^{n} \begin{bmatrix} n \\ 2k+1 \end{bmatrix}_q q^{\binom{n-(2k+1)}{2}} {}_c\mathcal{E}_{2k+1,q}(x,y) r^{n-(2k+1)}, & \text{if} \quad n=odd \\[2em] 2\sum_{k=0}^{n} \begin{bmatrix} n \\ 2k \end{bmatrix}_q q^{\binom{n-2k}{2}} {}_c\mathcal{E}_{2k,q}(x,y) r^{n-2k}, & \text{if} \quad n=even \end{cases} .$$

(ii) $\quad {}_s\mathcal{E}_{n,q}((x \oplus r)_q, y) + {}_s\mathcal{E}_{n,q}((x \ominus r)_q, y)$

$$= \begin{cases} 2\sum_{k=0}^{n} \begin{bmatrix} n \\ 2k+1 \end{bmatrix}_q q^{\binom{n-(2k+1)}{2}} {}_s\mathcal{E}_{2k+1,q}(x,y) r^{n-(2k+1)}, & \text{if} \quad n=odd \\[2em] 2\sum_{k=0}^{n} \begin{bmatrix} n \\ 2k \end{bmatrix}_q q^{\binom{n-2k}{2}} {}_s\mathcal{E}_{2k,q}(x,y) r^{n-2k}, & \text{if} \quad n=even \end{cases} .$$

(61)

Proof. (i) By using Lemma 1 (i) and (ii), we obtain

$$\begin{aligned} &{}_c\mathcal{E}_{n,q}((x \oplus r)_q, y) + {}_c\mathcal{E}_{n,q}((x \ominus r)_q, y) \\ &= \sum_{k=0}^{n} \begin{bmatrix} n \\ k \end{bmatrix}_q q^{\binom{n-k}{2}} \left({}_c\mathcal{E}_{k,q}(x,y) + (-1)^{n-k} {}_c\mathcal{E}_{k,q}(x,y) \right) r^{n-k}. \end{aligned} \tag{62}$$

If n is an odd or even number, then we derive the required result.

(ii) We omit the proof of Theorem 12 (ii) because we obtain the desired result in the same manner. $\square$

Corollary 7. *Let $r = 1$ in Theorem 12. Then, we have*

$$
\begin{aligned}
(i) \quad & {}_c\mathcal{E}_{n,q}((x \oplus 1)_q, y) + {}_c\mathcal{E}_{n,q}((x \ominus 1)_q, y) \\
&= \begin{cases}
2\sum_{k=0}^{n} \begin{bmatrix} n \\ 2k+1 \end{bmatrix}_q q^{\left(\frac{n-(2k+1)}{2}\right)} {}_c\mathcal{E}_{2k+1,q}(x,y), & \text{if} \quad n=\text{odd} \\[2em]
2\sum_{k=0}^{n} \begin{bmatrix} n \\ 2k \end{bmatrix}_q q^{\left(\frac{n-2k}{2}\right)} {}_c\mathcal{E}_{2k,q}(x,y), & \text{if} \quad n=\text{even}
\end{cases}
\end{aligned}
$$

$$
\begin{aligned}
(ii) \quad & {}_s\mathcal{E}_{n,q}((x \oplus 1)_q, y) + {}_s\mathcal{E}_{n,q}((x \ominus 1)_q, y) \\
&= \begin{cases}
2\sum_{k=0}^{n} \begin{bmatrix} n \\ 2k+1 \end{bmatrix}_q q^{\left(\frac{n-(2k+1)}{2}\right)} {}_s\mathcal{E}_{2k+1,q}(x,y), & \text{if} \quad n=\text{odd} \\[2em]
2\sum_{k=0}^{n} \begin{bmatrix} n \\ 2k \end{bmatrix}_q q^{\left(\frac{n-2k}{2}\right)} {}_s\mathcal{E}_{2k,q}(x,y), & \text{if} \quad n=\text{even}
\end{cases}
\end{aligned}
\tag{63}
$$

Corollary 8. *Let $q \to 1$ in Theorem 12. Then, we have*

$$
\begin{aligned}
(i) \quad & {}_c\mathcal{E}_n(x+r, y) + {}_c\mathcal{E}_n(x-r, y) \\
&= \begin{cases}
2\sum_{k=0}^{n} \binom{n}{2k+1} {}_c\mathcal{E}_{2k+1}(x,y) r^{n-(2k+1)}, & \text{if} \quad n=\text{odd} \\[1.5em]
2\sum_{k=0}^{n} \binom{n}{2k} {}_c\mathcal{E}_{2k}(x,y) r^{n-2k}, & \text{if} \quad n=\text{even}
\end{cases}
\end{aligned}
$$

$$
\begin{aligned}
(ii) \quad & {}_s\mathcal{E}_n(x+r, y) + {}_s\mathcal{E}_n(x-r, y) \\
&= \begin{cases}
2\sum_{k=0}^{n} \binom{n}{2k+1} {}_s\mathcal{E}_{2k+1}(x,y) r^{n-(2k+1)}, & \text{if} \quad n=\text{odd} \\[1.5em]
2\sum_{k=0}^{n} \binom{n}{2k} {}_s\mathcal{E}_{2k}(x,y) r^{n-2k}, & \text{if} \quad n=\text{even}
\end{cases}
\end{aligned}
\tag{64}
$$

Corollary 9. *From Lemma 1, one holds*

$$
\begin{aligned}
(i) \quad & {}_c\mathcal{E}_{n,q}((x \oplus r)_q, y) + {}_s\mathcal{E}_{n,q}((x \oplus r)_q, y) \\
&= \sum_{k=0}^{n} \begin{bmatrix} n \\ k \end{bmatrix}_q q^{\left(\frac{n-k}{2}\right)} \left({}_c\mathcal{E}_{k,q}(x,y) + {}_c\mathcal{E}_{k,q}(x,y) \right) r^{n-k},
\end{aligned}
$$

$$
\begin{aligned}
(ii) \quad & {}_c\mathcal{E}_{n,q}((x \ominus r)_q, y) + {}_s\mathcal{E}_{n,q}((x \ominus r)_q, y) \\
&= \sum_{k=0}^{n} \begin{bmatrix} n \\ k \end{bmatrix}_q (-1)^{n-k} q^{\left(\frac{n-k}{2}\right)} \left({}_c\mathcal{E}_{k,q}(x,y) + {}_c\mathcal{E}_{k,q}(x,y) \right) r^{n-k}.
\end{aligned}
\tag{65}
$$

Theorem 13. *For $|q| < 1$, we have the following relation:*

$$
\mathcal{E}_{n,q}(x) = \sum_{k=0}^{\left[\frac{n}{2}\right]} \begin{bmatrix} n \\ 2k \end{bmatrix}_q (-1)^k y^{2k} {}_c\mathcal{E}_{n-k,q}(x,y) + \sum_{k=0}^{\left[\frac{n-1}{2}\right]} \begin{bmatrix} n \\ 2k+1 \end{bmatrix}_q (-1)^k y^{2k+1} {}_s\mathcal{E}_{n-(2k+1),q}(x,y),
\tag{66}
$$

where $\mathcal{E}_{n,q}(x)$ is the q-Euler polynomials and $[x]$ is the greatest integer not exceeding x.

Proof. (i) We consider a multiplication between the generating function of q-cosine Euler polynomials and the q-cosine function such as

$$\sum_{n=0}^{\infty} {}_{C}\mathcal{E}_{n,q}(x,y)\frac{t^n}{[n]_q!}\cos_q(ty) = \frac{2}{e_q(t)+1}e_q(tx)COS_q(ty)\cos_q(ty). \tag{67}$$

By using the power series of a q-cosine function, the left-hand side of Equation (67) is written as

$$
\begin{aligned}
\sum_{n=0}^{\infty} {}_{C}\mathcal{E}_{n,q}(x,y)\frac{t^n}{[n]_q!}\cos_q(ty) &= \sum_{n=0}^{\infty} {}_{C}\mathcal{E}_{n,q}(x,y)\frac{t^n}{[n]_q!}\sum_{n=0}^{\infty}(-1)^n y^{2n}\frac{t^{2n}}{[2n]_q!}\\
&= \sum_{n=0}^{\infty}\left(\sum_{k=0}^{n}\begin{bmatrix}n+k\\2k\end{bmatrix}_q(-1)^k y^{2k}{}_{C}\mathcal{E}_{n-k,q}(x,y)\right)\frac{t^{n+k}}{[n+k]_q!}\\
&= \sum_{n=0}^{\infty}\left(\sum_{k=0}^{[\frac{n}{2}]}\begin{bmatrix}n\\2k\end{bmatrix}_q(-1)^k y^{2k}{}_{C}\mathcal{E}_{n-k,q}(x,y)\right)\frac{t^n}{[n]_q!}.
\end{aligned}
\tag{68}
$$

From Equations (67) and (68), we have

$$\sum_{n=0}^{\infty}\left(\sum_{k=0}^{[\frac{n}{2}]}\begin{bmatrix}n\\2k\end{bmatrix}_q(-1)^k y^{2k}{}_{C}\mathcal{E}_{n-k,q}(x,y)\right)\frac{t^n}{[n]_q!} = \frac{2}{e_q(t)+1}e_q(tx)COS_q(ty)\cos_q(ty). \tag{69}$$

In a similar way, we find the multiplication between the q-sin Euler polynomials and the q-sin function as follows.

$$\sum_{n=0}^{\infty} {}_{S}\mathcal{E}_{n,q}(x,y)\frac{t^n}{[n]_q!}\sin_q(ty) = \frac{2}{e_q(t)+1}e_q(tx)SIN_q(ty)\sin_q(ty). \tag{70}$$

Applying the power series of a q-sine function, the left-hand side of Equation (70) is obtained as

$$
\begin{aligned}
\sum_{n=0}^{\infty} {}_{S}\mathcal{E}_{n,q}(x,y)\frac{t^n}{[n]_q!}\sin_q(ty) &= \sum_{n=0}^{\infty} {}_{S}\mathcal{E}_{n,q}(x,y)\frac{t^n}{[n]_q!}\sum_{n=0}^{\infty}(-1)^n y^{2n+1}\frac{t^{2n+1}}{[2n+1]_q!}\\
&= \sum_{n=0}^{\infty}\left(\sum_{k=0}^{[\frac{n-1}{2}]}\begin{bmatrix}n\\2k+1\end{bmatrix}_q(-1)^k y^{2k+1}{}_{S}\mathcal{E}_{n-(2k+1),q}(x,y)\right)\frac{t^n}{[n]_q!}.
\end{aligned}
\tag{71}
$$

We can find (72) by using $\cos_q(ty)COS_q(ty) + \sin_q(ty)SIN_q(ty) = 1$, which is a property of q-trigonometric functions.

$$
\begin{aligned}
&\sum_{n=0}^{\infty}\left(\sum_{k=0}^{[\frac{n}{2}]}\begin{bmatrix}n\\2k\end{bmatrix}_q(-1)^k y^{2k}{}_{C}\mathcal{E}_{n-k,q}(x,y) + \sum_{k=0}^{[\frac{n-1}{2}]}\begin{bmatrix}n\\2k+1\end{bmatrix}_q(-1)^k y^{2k+1}{}_{S}\mathcal{E}_{n-(2k+1),q}(x,y)\right)\frac{t^n}{[n]_q!}\\
&= \frac{2}{e_q(t)+1}e_q(tx)\left(\cos_q(ty)COS_q(ty) + \sin_q(ty)SIN_q(ty)\right)\\
&= \sum_{n=0}^{\infty}\mathcal{E}_{n,q}(x)\frac{t^n}{[n]_q!}.
\end{aligned}
\tag{72}
$$

From Equation (72), we can find the required result of Theorem 13. $\square$

Corollary 10. *If $q \to 1$ in Theorem 13, then we have*

$$\mathcal{E}_n(x) = \sum_{k=0}^{\left[\frac{n}{2}\right]} \binom{n}{2k}(-1)^k y^{2k}{}_C\mathcal{E}_{n-k}(x,y) + \sum_{k=0}^{\left[\frac{n-1}{2}\right]} \binom{n}{2k+1}(-1)^k y^{2k+1}{}_S\mathcal{E}_{n-(2k+1)}(x,y). \tag{73}$$

Corollary 11. *Setting $y = 1$ in Theorem 13, one holds*

$$\mathcal{E}_{n,q}(x) = \sum_{k=0}^{\left[\frac{n}{2}\right]} \begin{bmatrix} n \\ 2k \end{bmatrix}_q (-1)^k{}_C\mathcal{E}_{n-k,q}(x,1) + \sum_{k=0}^{\left[\frac{n-1}{2}\right]} \begin{bmatrix} n \\ 2k+1 \end{bmatrix}_q (-1)^k{}_S\mathcal{E}_{n-(2k+1),q}(x,1). \tag{74}$$

Corollary 12. *From the Theorem 13 and Corollary 11, we have*

$$\sum_{k=0}^{\left[\frac{n}{2}\right]} \begin{bmatrix} n \\ 2k \end{bmatrix}_q (-1)^k \left(y^{2k}{}_C\mathcal{E}_{n-k,q}(x,y) - {}_C\mathcal{E}_{n-k,q}(x,1)\right)$$

$$= \sum_{k=0}^{\left[\frac{n-1}{2}\right]} \begin{bmatrix} n \\ 2k+1 \end{bmatrix}_q (-1)^k \left({}_S\mathcal{E}_{n-(2k+1),q}(x,1) - y^{2k+1}{}_S\mathcal{E}_{n-(2k+1),q}(x,y)\right). \tag{75}$$

To find a relationship between the q-cosine Euler polynomials and the q-cosine Bernoulli polynomials, we recall the definitions of q-cosine and q-sine Bernoulli polynomials, see [15]. The q-cosine Bernoulli polynomials ${}_C B_n(x,y)$ and q-cosine Bernoulli polynomials ${}_S B_n(x,y)$ are defined by means of the generating functions

$$\sum_{n=0}^{\infty} {}_C B_{n,q}(x,y)\frac{t^n}{n!} = \frac{t}{e_q(t)-1}e_q(tx)\text{COS}_q(ty),$$

$$\sum_{n=0}^{\infty} {}_S B_{n,q}(x,y)\frac{t^n}{n!} = \frac{t}{e_q(t)-1}e_q(tx)\text{SIN}_q(ty). \tag{76}$$

Theorem 14. *Let $x, y \in \mathbb{R}$ and $|q| < 1$. Then we derive*

$$\begin{aligned}
(i) \quad & [n]_q{}_C\mathcal{E}_{n-1,q}(x,y) + 2{}_C B_{n,q}(x,y) \\
& = \sum_{k=0}^{n} \begin{bmatrix} n \\ k \end{bmatrix}_q \left(2{}_C B_{k,q}(x,y) - [k]_q{}_C\mathcal{E}_{k-1,q}(x,y)\right), \\
(ii) \quad & [n]_q{}_S\mathcal{E}_{n-1,q}(x,y) + 2{}_S B_{n,q}(x,y) \\
& = \sum_{k=0}^{n} \begin{bmatrix} n \\ k \end{bmatrix}_q \left(2{}_S B_{k,q}(x,y) - [k]_q{}_S\mathcal{E}_{k-1,q}(x,y)\right),
\end{aligned} \tag{77}$$

where ${}_C B_{n,q}(x,y)$ is the q-cosine Bernoulli polynomials and ${}_S B_{n,q}(x,y)$ is the q-sine Bernoulli polynomials.

Proof. (i) We substitute the generating function of q-cosine Euler polynomials with an expression that is related to the q-cosine Bernoulli polynomials as

$$\sum_{n=0}^{\infty} {}_C\mathcal{E}_{n,q}(x,y)\frac{t^n}{[n]_q!} = \frac{2(e_q(t)-1)}{t(e_q(t)+1)} \sum_{n=0}^{\infty} {}_C B_{n,q}(x,y)\frac{t^n}{[n]_q!}. \tag{78}$$

From Equation (78), we have

$$\sum_{n=0}^{\infty} {}_C\mathcal{E}_{n,q}(x,y)\frac{t^{n+1}}{[n]_q!}\left(\sum_{n=0}^{\infty} \frac{t^n}{[n]_q!} + 1\right) = 2\sum_{n=0}^{\infty} {}_C B_{n,q}(x,y)\frac{t^n}{[n]_q!}\left(\sum_{n=0}^{\infty} \frac{t^n}{[n]_q!} - 1\right). \tag{79}$$

We replace the left-hand side of (79) with the following equation.

$$
\sum_{n=0}^{\infty} {}_C\mathcal{E}_{n,q}(x,y)\frac{t^{n+1}}{[n]_q!}\left(\sum_{n=0}^{\infty}\frac{t^n}{[n]_q!}+1\right)
$$

$$
= \sum_{n=0}^{\infty} [n]_q {}_C\mathcal{E}_{n-1,q}(x,y)\frac{t^n}{[n]_q!}\left(\sum_{n=0}^{\infty}\frac{t^n}{[n]_q!}+1\right) \tag{80}
$$

$$
= \sum_{n=0}^{\infty}\left(\sum_{k=0}^{n}\begin{bmatrix}n\\k\end{bmatrix}_q [k]_q {}_C\mathcal{E}_{k-1,q}(x,y)+[n]_q {}_C\mathcal{E}_{n-1,q}(x,y)\right)\frac{t^n}{[n]_q!}.
$$

Then, the right-hand side of (79) is transformed as

$$
2\sum_{n=0}^{\infty} {}_C B_{n,q}(x,y)\frac{t^n}{[n]_q!}\left(\sum_{n=0}^{\infty}\frac{t^n}{[n]_q!}-1\right)=2\sum_{n=0}^{\infty}\left(\sum_{k=0}^{n}\begin{bmatrix}n\\k\end{bmatrix}_q {}_C B_{k,q}(x,y)-{}_C B_{n,q}(x,y)\right)\frac{t^n}{[n]_q!}. \tag{81}
$$

By comparing Equations (80) and (81), we investigate a relation between the q-cosine Euler polynomials and q-cosine Bernoulli polynomials and complete the proof of Theorem 14.

(ii) By using a similar method as in (i), we derive the required result. $\square$

Corollary 13. *When $q \to 1$ in Theorem 14, the following holds*

$$
\begin{aligned}
(i)\quad & n_C\mathcal{E}_{n-1}(x,y)+2_C B_n(x,y) = \sum_{k=0}^{n}\binom{n}{k}\left(2_C B_k(x,y)-k_C\mathcal{E}_{k-1}(x,y)\right),\\
(ii)\quad & n_S\mathcal{E}_{n-1}(x,y)+2_S B_n(x,y) = \sum_{k=0}^{n}\binom{n}{k}\left(2_S B_k(x,y)-k_S\mathcal{E}_{k-1}(x,y)\right),
\end{aligned} \tag{82}
$$

where $_C B_n(x,y)$ is the cosine Bernoulli polynomials, and $_S B_n(x,y)$ is the sine Bernoulli polynomials.

4. Symmetric Structures of Approximate Roots for q-cosine Euler Polynomials and Their Application

In this section, we show the actual forms for q-cosine and q-sine Euler polynomials using the theorems from Section 2 and the Mathematica program. We observe the structure of the approximate roots of these polynomials and find some properties. We also show examples of q-cosine Euler polynomials using Newton's method.

First, we discuss q-cosine Euler polynomials. A few forms of q-cosine Euler polynomials are as follows:

$$c\mathcal{E}_{0,q}(x,y) = 1,$$

$$c\mathcal{E}_{1,q}(x,y) = -\frac{1}{2} + x,$$

$$c\mathcal{E}_{2,q}(x,y) = \frac{1}{4}(-1 + q - 2x - 2qx + 4x^2 - 4qy^2),$$

$$c\mathcal{E}_{3,q}(x,y) = \frac{1}{8}(-1 + 2q + 2q^2 - q^3 - 2x + 2q^3x - 4x^2 - 4qx^2 - 4q^2x^2 + 8x^3)$$
$$- \frac{1}{8}(4q(1 + q + q^2)(-1 + 2x)y^2),$$

$$c\mathcal{E}_{4,q}(x,y) = \frac{1}{16}(-1 + 3q + 3q^2 - 3q^4 - 3q^5 + q^6 - 2x + 2qx + 6q^2x + 4q^3x + 6q^4x)$$
$$+ \frac{1}{16}(2q^5x - 2q^6x - 4x^2 - 4q^2x^2 + 4q^3x^2 + 4q^5x^2 - 8x^3 - 8qx^3 - 8q^2x^3)$$
$$- \frac{1}{16}(8q^3x^3 - 16x^4 + 4q(1 + q^2)(1 + q + q^2)(1 - q + 2(1 + q)x - 4x^2)y^2)$$
$$+ q^6y^4,$$

$$\cdots .$$

Next, we show the approximate roots table of q-cosine Euler polynomials. Based on Equation (83), we construct Table 1 for the approximate roots of q-cosine Euler polynomials. In Table 1, we vary the values of p and n when $y = 7$. Then, we obtain only real roots with $1 \le n \le 7$ in $q = 0.5$ and $q = 0.9$. From Table 1, we can consider two previews:

(1) When n increases, the absolute values of the real roots approach to approximately 2.345 and 1 for $q = 0.1$.

(2) When q approaches 1, the approximate root distribution of q-cosine Euler polynomials are spread and most of them appear as real roots.

Figure 1 shows the structure of the approximate roots for q-cosine Euler polynomials. Let $y = 7$ and $1 \le n \le 30$. The left graph in Figure 1 is $q=0.99$, the middle graph is $q = 0.9$, and the right graph is $q = 0.8$. The blue color denotes that n is small, and the red color denotes that n is 30. In Figure 1, we show that the approximate roots of q-cosine Euler polynomials include all real numbers in $q = 0.99$ when $n = 30$. In addition, we conject that the approximate roots of q-cosine Euler polynomials show a circle structure near 0 when q approaches 0 and n continues to increase.

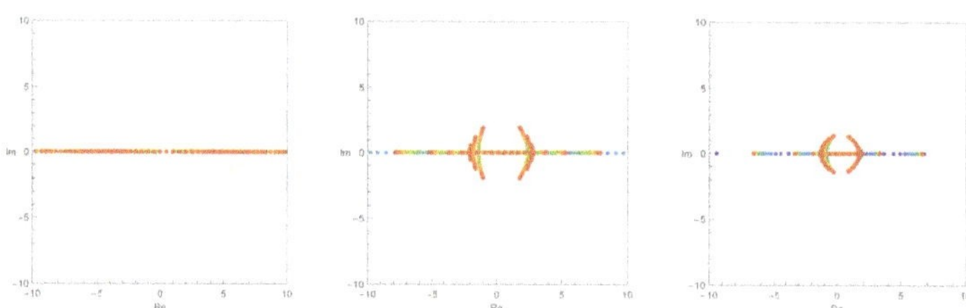

Figure 1. Stacking structure of approximation roots in q-cosine Euler polynomials when $q = 0.99, 0.9,$ and 0.8 and $y = 7$.

Table 1. Approximate zeros of $\mathcal{E}_{n,q}(x,7)$.

n	$q = 0.1$	$q = 0.5$	$q = 0.9$
1	0.5	0.5	0.5
2	-2.00549, 2.55549	-4.60151, 5.35151	-6.18463, 7.13463
3	-2.3461, 0.457296, 2.44381	-6.3923, 0.496385, 6.77091	-10.5354, 0.499698, 11.3907
4	$-2.33937, -0.271276$, 0.795233, 2.37091	$-7.20113, -0.610105$, 1.36484, 7.3839	$-14.1519, -2.0322$, 2.98714, 14.9165
5	-2.34547, $-0.185877 - 0.259313i$, $-0.185877 + 0.259313i$, 0.92226, 2.35051	$-7.58727, -1.07425$, 0.403015, 1.55082, 7.67644	$-17.2997, -3.75091$, 0.497805, 4.617, 17.9833
6	$-2.34484, -0.335526$, $-0.0397254 - 0.350112i$, $-0.0397254 + 0.350112i$, 0.969307, 2.34607	$-7.77643, -1.22901$, $-0.200224, 0.852848$, 1.51678, 7.82041	$-20.0834, -5.08688$, $-1.00794, 1.96854$, 5.85703, 20.6955
7	-2.345, $-0.293507 - 0.196006i$, $-0.293507 + 0.196006i$, $0.0774528 - 0.366718i$, $0.0774528 + 0.366718i$, 0.98748, 2.34519	$-7.87012, -1.28288$, $-0.40639, 0.14643$, 1.10906, 1.40412, 7.89196	$-22.5628, -6.18568$, -2.10138, 0.493322, 2.98756, 6.86617, 23.1113
$\vdots$	$\vdots$	$\vdots$	$\vdots$

Figure 2 shows the 3D structure of Figure 1 under the same conditions. The left shape is the approximate roots of q-cosine Euler polynomials when $q = 0.99$, $y = 7$, and $1 \le n \le 30$. This shape indicates that all the approximate roots are located on an imaginary axis. The middle shape in Figure 2 shows the approximate roots of q-cosine Euler polynomials when $q = 0.9$, $y = 7$, and $1 \le n \le 30$. Here, we can observe the movement of the approximate roots. When $q = 0.8$ and $y = 7$, we can see the right shape of Figure 2. The shape variation in Figure 2 implies that the approximate roots change to imaginary numbers from real numbers and that the root structure for q-cosine Euler polynomials vary according to q.

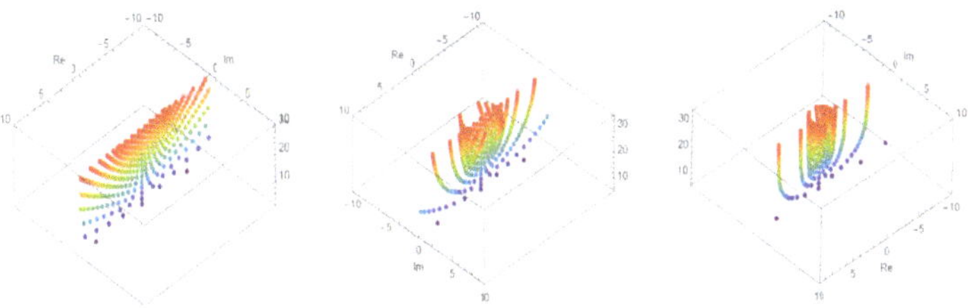

Figure 2. Stacking structure of the approximation roots in q-cosine Euler polynomials when $q = 0.99, 0.9$, and 0.8 and $y = 7$ in 3D.

Conjecture 1. *If n increases and $q \to 0$, then the approximate roots of q-cosine Euler polynomials display a circle shape near the origin except for some zeros.*

In Figure 3, $q = 0.99$ and $y = 7$ when $1 \leq n \leq 30$. Under these conditions, we observe that the approximate roots of q-cosine Euler polynomials have a symmetric property and include all real numbers. By observing the right graphs in Figures 1 and 3, we can consider Conjecture 2.

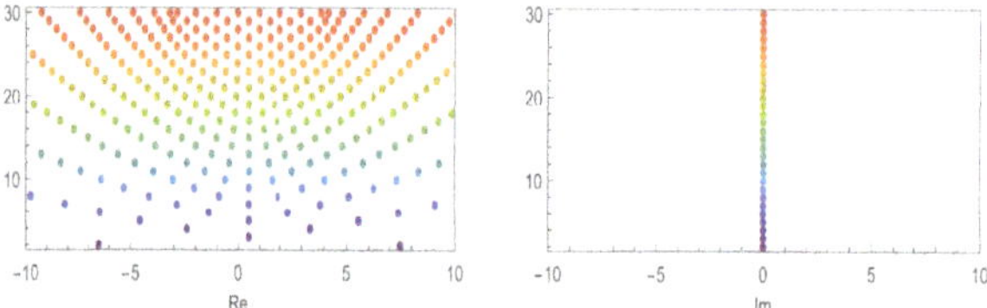

Figure 3. Stacking structure of the approximate roots in q-cosine Euler polynomials when $q = 0.99$ and $y = 7$.

Conjecture 2. *Prove that* $_c\mathcal{E}_{n,q}(x,y)$ *is reflection symmetry analytic complex functions which has* $Re(x) = 1/2$ *in addition to the usual* $Im(x) = 0$, *when y is a fixed point in real numbers.*

By using the Newton's method(see [22]), we see the following Example 1. The equation of the left figure in Example 3 is $1.1965181875000004 + 2.912903125000001x - 5.745637500000002x^2 - 0.5555x^3 + x^4$, that is $q = 0.1$. In Table 1, we note that the approximate roots are $-2.33937, -0.271276, 0.795233$, and 2.37091, where $q = 0.1$ and $y = 7$. When we choose $-4 \leq Re(x) \leq 4$ and $-4 \leq Im(x) \leq 4$, we obtain the left figure in a complex plane. The complex numbers in the red, violet , yellow, and sky-blue ranges move to $-2.3397, -0.271276, 0.795233$, and 2.37091, respectively. The right figure in Example 3 illustrates the 4-th q-cosine Euler polynomials when $q = 0.5$ and $y = 7$. Numbers of the red, violet, yellow, and sky-blue ranges in the complex plane become $-7.20113, -0.610105, 1.36484$, and 7.3839, respectively (Figure 4).

Example 3. *The 4-th q-cosine Euler polynomials display the following figures in a complex plane:*

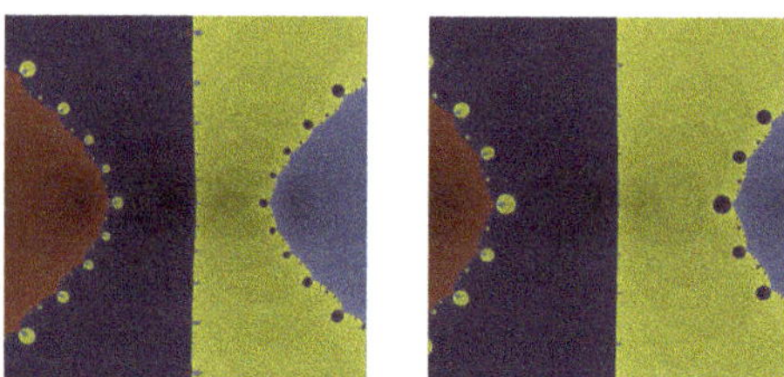

Figure 4. The 4-th q-cosine Euler polynomials for $q = 0.1$ and $y = 7$.

5. Conclusions

In this paper, we have identified several properties of q-cosine Euler polynomials and q-sine Euler polynomials. In addition, the relationship between polynomials was confirmed according to the various conditions of the variables. We were able to assume the structure of the approximate roots of the q-cosine Euler polynomials and the q-sine Euler polynomials and finally, produce some speculations. The structure of the approximate roots will come out in various ways depending on the condition of the variables, and new methods and theorems related to approaching this needs to be created and proved.

Author Contributions: Conceptualization, J.Y.K.; Data curation, C.S.R.; Formal analysis, C.S.R.; Methodology, J.Y.K.; Software, J.Y.K.; Writing—original draft, J.Y.K. These authors contributed equally to this work. All authors have read and agreed to the published version of the manuscript.

Funding: This work was supported by the Basic Science Research Program through the National Research Foundation of Korea (NRF) funded by the Ministry of Science, ICT and Future Planning (No. 2017R1E1A1A03070483).

Acknowledgments: The authors would like to thank the referees for their valuable comments, which improved the original manuscript in its present form.

Conflicts of Interest: The authors declare that there is no conflict of interests regarding the publication of this paper.

References

1. Jackson, H.F. q-Difference equations. *Am. J. Math.* **1910**, *32*, 305–314. [CrossRef]
2. Charalambides, C.A. *Discrete q-Distributions*; Wiley: New York, NY, USA, 2016; ISBN 9781119119050.
3. Ernst, T. *A Comprehensive Treatment of q-Calculus*; Springer Science & Business Media: New York, NY, USA, 2012.
4. Gasper, G.; Rahman, M. *Basic Hypergeometric Series*, 2nd ed.; Cambridge University Press: Cambridge, UK, 2004; ISBN: 9780511526251.
5. Andrews, G.E.; Crippa, D.; Simon, K. q-Series arising from the study of random graphs. *SIAM J. Discrete Math.* **1997**, *10*, 41–56. [CrossRef]
6. Charalambides, C.A. On the q-differences of the generalized q-factorials. *J. Statist. Plann. Inference* **1996**, *54*, 31–43. [CrossRef]
7. Exton, H. *q-hypergeometric Functions and Applications*; Halstead Press: New York, NY, USA; Ellis Horwood: Chichester, UK, 1983; ISBN 0853124914.
8. Kemp, A. Certain q-analogues of the binomial distribution. *Sankhya A* **2002**, *64*, 293–305.
9. Ostrovska, S. q-Bernstein polynomials and their iterates. *J. Approx. Theory* **2003**, *123*, 232–255. [CrossRef]
10. Rawlings, D. Limit formulas for q-exponential functions. *Discrete Math.* **1994**, *126*, 379–383. [CrossRef]
11. Sparavigna, A.C. Graphs of q-Exponentials and q-Trigonometric Functions. 2016. Available online: https://hal.archives-ouvertes.fr/hal-01377262 (accessed on 15 June 2020).
12. Jackson, H.F. On q-functions and a certain difference operator. *Trans. R. Soc. Edinb.* **2013**, *46*, 253–281. [CrossRef]
13. Phillips, G.M. Bernstein polynomials based on q-integers. *Ann. Numer. Anal.* **1997**, *4*, 511–518.
14. Kac, V.; Pokman, C. *Quantum Calculus Universitext*; Springer: New York, NY, USA, 2002; ISBN 9781461300717.
15. Kang, J.Y.; Ryoo, C.S. Various structures of the roots and explicit properties of q-cosine Bernoulli polynomials and q-sine Bernoulli polynomials. *Mathematics* **2020**, *8*, 463. [CrossRef]
16. Kim, T.; Ryoo, C.S. Some identities for Euler and Bernoulli polynomials and their zeros. *Axioms* **2018**, *7*, 56. [CrossRef]
17. Kim, T.; Ryoo, C.S. Sheffer Type Degenerate Euler and Bernoulli Polynomials. *Filomat* **2019**, *33*, 6173–6185. [CrossRef]
18. Liu, H.; Wang, W. Some identities on the Bernoulli, Euler and Genocchi polynomials via power sums and alternate power sums. *Discrete Math.* **2009**, *309*, 3346–3363. [CrossRef]
19. Mahmudov, N.I.; Momenzadeh, M. On a class of q-Bernoulli, q-Euler, and q-Genocchi polynomials. *Abstr. Appl. Anal.* **2014**, *2014*, 696454. [CrossRef]
20. Park, M.J.; Kang, J.Y. A study on the cosine tangent polynomials and sine tangent polynomials. *J. Appl. Pure Math.* **2020**, *2*, 47–56.
21. Carlitz, L. Degenerate Stiling, Bernoulli and Eulerian numbers. *Utilitas Math.* **1979**, *15*, 51–88.
22. Kharab, A.; Guenther, R.B. *An Introduction to Numerical Methods: A Matlab Approach*, 3rd ed.; Chapman and Hall/crc: Boca Raton, FL, USA, 2018; ISBN-13 978-1439868997.

© 2020 by the authors. Licensee MDPI, Basel, Switzerland. This article is an open access article distributed under the terms and conditions of the Creative Commons Attribution (CC BY) license (http://creativecommons.org/licenses/by/4.0/).

Article

Certain Results for the Twice-Iterated 2D q-Appell Polynomials

Hari M. Srivastava [1,2,*], Ghazala Yasmin [3], Abdulghani Muhyi [3] and Serkan Araci [4]

[1] Department of Mathematics and Statistics, University of Victoria, Victoria, BC V8W 3R4, Canada
[2] Department of Medical Research, China Medical University Hospital, China Medical University, Taichung 40402, Taiwan
[3] Department of Applied Mathematics, Aligarh Muslim University, Aligarh 202002, Uttar Pradesh, India; ghazala30@gmail.com (G.Y.); muhyi2007@gmail.com (A.M.)
[4] Department of Economics, Faculty of Economics, Administrative and Social Sciences, Hasan Kalyoncu University, TR-27410 Gaziantep, Turkey; serkan.araci@hku.edu.tr
* Correspondence: harimsri@math.uvic.ca; Tel.: +1-250-472-5313 (Office); +1-250-477-6960 (Home)

Received: 16 September 2019; Accepted: 12 October 2019; Published: 16 October 2019

Abstract: In this paper, the class of the twice-iterated 2D q-Appell polynomials is introduced. The generating function, series definition and some relations including the recurrence relations and partial q-difference equations of this polynomial class are established. The determinant expression for the twice-iterated 2D q-Appell polynomials is also derived. Further, certain twice-iterated 2D q-Appell and mixed type special q-polynomials are considered as members of this polynomial class. The determinant expressions and some other properties of these associated members are also obtained. The graphs and surface plots of some twice-iterated 2D q-Appell and mixed type 2D q-Appell polynomials are presented for different values of indices by using Matlab. Moreover, some areas of potential applications of the subject matter of, and the results derived in, this paper are indicated.

Keywords: 2D q-Appell polynomials; twice-iterated 2D q-Appell polynomials; determinant expressions; recurrence relations; 2D q-Bernoulli polynomials; 2D q-Euler polynomials; 2D q-Genocchi polynomials; Apostol type Bernoulli; Euler and Genocchi polynomials

PACS: 33D45; 33D99; 33E20

---

1. Introduction, Definitions and Preliminaries

The subject of q-calculus leads to a new method for computations and classifications of q-series and q-polynomials. In fact, the subject of q-calculus was initiated in the 1920s. However, it has gained considerable popularity and importance during the last three decades or so. In the past decade, q-calculus was developed into an interdisciplinary subject and it served as a bridge between mathematics and physics. The field has been expanded explosively due mainly to its applications in diverse areas of physics such as cosmic strings and black holes [1], conformal quantum mechanics [2], nuclear and high energy physics [3], fluid mechanics, combinatorics, having connection with commutativity relations, number theory, and Lie algebra. The definitions and notations of the q-calculus reviewed here are taken from [4] (see also [5,6]).

The q-analogue of the Pochhammer symbol $(\alpha)_m$, which is also called the q-shifted factorial, defined by

$$(\alpha;q)_0 = 1 \quad \text{and} \quad (\alpha;q)_m = \prod_{r=0}^{m-1}(1 - \alpha q^r) \qquad (m \in \mathbb{N};\ \alpha \in \mathbb{C}). \tag{1}$$

The q-analogues of a complex number α and of the factorial function are defined as follows:

$$[\alpha]_q = \frac{1 - q^\alpha}{1 - q} \qquad (q \in \mathbb{C} \setminus \{1\}; \ \alpha \in \mathbb{C}) \tag{2}$$

and

$$[m]_q = \sum_{s=1}^{m} q^{s-1}, \ [0]_q = 0, \ [m]_q! = \prod_{s=1}^{m} [s]_q = [1]_q [2]_q [3]_q \cdots [m]_q \ \text{ and } \ [0]_q! = 1 \tag{3}$$

$$(m \in \mathbb{N}; \ q \in \mathbb{C} \setminus \{0, 1\}),$$

where $\mathbb{N}$ is the set of positive integers.

The q-binomial coefficients $\begin{bmatrix} m \\ s \end{bmatrix}_q$ are defined by

$$\begin{bmatrix} m \\ s \end{bmatrix}_q = \frac{(q;q)_m}{(q;q)_s (q;q)_{m-s}} = \frac{[m]_q!}{[s]_q! \, [m-s]_q!} \qquad (s = 0, 1, 2, \cdots, m). \tag{4}$$

The q-analogue of the classical derivative Df or $\frac{d}{dt} f$ of a function f at a point $t \in \mathbb{C} \setminus \{0\}$ is defined by

$$D_q f(t) = \frac{f(t) - f(qt)}{(1 - q)t} \qquad (0 < |q| < 1; \ t \neq 0). \tag{5}$$

We also note that

(i) $\quad \lim\limits_{q \to 1} D_q f(t) = \dfrac{df(t)}{dt}$, where $\dfrac{d}{dt}$ denotes the classical ordinary derivative,

(ii) $\quad D_q \big(a_1 f(t) + a_2 \, g(t)\big) = a_1 D_q f(t) + a_2 D_q g(t)$,

(iii) $\quad D_q (fg)(t) = f(qt) D_q g(t) + g(t) D_q f(t) = f(t) D_q g(t) + D_q f(t) g(qt)$,

(vi) $\quad D_q \left(\dfrac{f(t)}{g(t)} \right) = \dfrac{g(t) D_q f(t) - f(t) D_q g(t)}{g(t) g(qt)} = \dfrac{g(qt) D_q f(t) - f(qt) D_q g(t)}{g(t) g(qt)}.$

The two familiar q-analogues of the exponential function e^t are given by

$$e_q(t) := \sum_{m=0}^{\infty} \frac{t^m}{[m]_q!} = \frac{1}{\big((1-q)x; q\big)_\infty}, \quad 0 < |q| < 1, |x| < |1 - q|^{-1} \tag{6}$$

and

$$E_q(t) := \sum_{m=0}^{\infty} q^{\frac{1}{2}m(m-1)} \frac{t^m}{[m]_q!} = (-(1-q); q)_\infty \qquad (0 < |q| < 1; \ t \in \mathbb{C}). \tag{7}$$

The above-defined q-exponential functions $e_q(t)$ and $E_q(t)$ satisfy the following properties:

$$D_q e_q(t) = e_q(t), \ D_q E_q(t) = E_q(qt), \tag{8}$$

$$e_q(t) E_q(-t) = E_q(t) e_q(-t) = 1. \tag{9}$$

The class of Appell polynomials was introduced and characterized completely by Appell [7] in 1880. Further, Throne [8], Sheffer [9] and Varma [10] studied this class of polynomials from different points of view. For some subsequent and recent developments associated with the Appell polynomials, one may refer to the works [11–14].

In the year 1954, Sharma and Chak [15] introduced a q-analogue of the Appell polynomials and called this sequence of polynomials as q-Harmonics. Later, in the year 1967, Al-Salam [16] introduced the class of the q-Appell polynomials $\{\mathcal{A}_{m,q}(x)\}_{m=0}^{\infty}$ and studied some of their properties. Some characterizations of the q-Appell polynomials were presented by Srivastava [17] in the year 1982. These polynomials arise in numerous problems of applied mathematics, theoretical physics, approximation theory and many other branches of the mathematical sciences [7,18–20]. The polynomials $\mathcal{A}_{m,q}(x)$ (of degree m) are called q-Appell polynomials, provided that they satisfy the following q-differential equation:

$$D_{q,x}\{\mathcal{A}_{m,q}(x)\} = [m]_q \mathcal{A}_{m-1,q}(x) \qquad (m \in \mathbb{N}_0 = \mathbb{N} \cup \{0\}; \; q \in \mathbb{C}; \; 0 < |q| < 1). \tag{10}$$

Recently, Keleshteri and Mahmudov [21] introduced the 2D q-Appell polynomials (2DqAP) $\{\mathcal{A}_{m,q}(x_1, x_2)\}_{m=0}^{\infty}$ which are defined by means of the following generating function:

$$\mathcal{A}_q(t) \, e_q(x_1 t) E_q(x_2 t) = \sum_{m=0}^{\infty} \mathcal{A}_{m,q}(x_1, x_2) \frac{t^m}{[m]_q!} \qquad (0 < q < 1), \tag{11}$$

where

$$\mathcal{A}_q(t) = \sum_{m=0}^{\infty} \mathcal{A}_{m,q} \frac{t^m}{[m]_q!}, \quad \mathcal{A}_q(t) \neq 0 \quad \text{and} \quad \mathcal{A}_{0,q} = 1. \tag{12}$$

We write

$$\mathcal{A}_{m,q} := \mathcal{A}_{m,q}(0,0),$$

where $\mathcal{A}_{m,q}$ denotes the 2D q-Appell numbers.

For $x_2 = 0$, the 2DqAP $\mathcal{A}_{m,q}(x_1, x_2)$ reduce to the q-Appell polynomials $\mathcal{A}_{m,q}(x)$ (see, for example, [16,17,22]), that is,

$$\mathcal{A}_{m,q}(x_1, 0) = \mathcal{A}_{m,q}(x_1), \tag{13}$$

where $\mathcal{A}_{m,q}(x)$ are defined by

$$\mathcal{A}_q(t) \, e_q(xt) = \sum_{m=0}^{\infty} \mathcal{A}_{m,q}(x) \frac{t^m}{[m]_q!} \qquad (0 < q < 1) \tag{14}$$

and $\mathcal{A}_{m,q}$ given by

$$\mathcal{A}_{m,q} := \mathcal{A}_{m,q}(0)$$

denotes the q-Appell numbers.

The explicit form of the 2DqAP $\mathcal{A}_{m,q}(x_1, x_2)$ in terms qAP $\mathcal{A}_{m,q}(x)$ is given as follows (see [21]):

$$\mathcal{A}_{m,q}(x_1, x_2) = \sum_{s=0}^{m} \begin{bmatrix} m \\ s \end{bmatrix}_q q^{\frac{1}{2}(m-s)(m-s-1)} \mathcal{A}_{s,q}(x_1) x_2^{m-s}. \tag{15}$$

The function $\mathcal{A}_q(t)$ may be called the determining function for the set $\mathcal{A}_{m,q}(x_1, x_2)$. Based on suitable selections for the function $\mathcal{A}_q(t)$, the following different members belonging to the family of the 2D q-Appell polynomials $\mathcal{A}_{m,q}(x_1, x_2)$ can be obtained:

I. If $\mathcal{A}_q(t) = \frac{t}{e_q(t)-1}$, the 2DqAP $\mathcal{A}_{m,q}(x_1, x_2)$ reduce to the 2D q-Bernoulli polynomials (2DqBP) $\mathcal{B}_{m,q}(x_1, x_2)$ (see [23,24]), that is,

$$\mathcal{A}_{m,q}(x_1, x_2) = \mathcal{B}_{m,q}(x_1, x_2),$$

where $\mathfrak{B}_{m,q}(x_1, x_2)$ are defined by

$$\frac{t}{e_q(t) - 1} e_q(x_1 t) E_q(x_2 t) = \sum_{m=0}^{\infty} \mathfrak{B}_{m,q}(x_1, x_2) \frac{t^m}{[m]_q!} \tag{16}$$

and $\mathfrak{B}_{m,q}$ given by

$$\mathfrak{B}_{m,q} := \mathfrak{B}_{m,q}(0,0)$$

denotes the 2D q-Bernoulli numbers.

II. If $\mathcal{A}_q(t) = \frac{2}{e_q(t)+1}$, the 2DqAP $\mathcal{A}_{m,q}(x_1, x_2)$ reduce to the 2D q-Euler polynomials (2DqEP) $\mathcal{E}_{m,q}(x_1, x_2)$ (see [23,24]), that is,

$$\mathcal{A}_{m,q}(x_1, x_2) = \mathcal{E}_{m,q}(x_1, x_2),$$

where $\mathcal{E}_{m,q}(x_1, x_2)$ are defined by

$$\frac{2}{e_q(t) + 1} e_q(x_1 t) E_q(x_2 t) = \sum_{m=0}^{\infty} \mathcal{E}_{m,q}(x_1, x_2) \frac{t^m}{[m]_q!} \tag{17}$$

and $\mathcal{E}_{m,q}$ given by

$$\mathcal{E}_{m,q} := \mathcal{E}_{m,q}(0,0)$$

denotes the 2D q-Euler numbers.

III. If $\mathcal{A}_q(t) = \frac{2t}{e_q(t)+1}$, the 2DqAP $\mathcal{A}_{m,q}(x_1, x_2)$ reduce to the 2D q-Genocchi polynomials (2DqGP) $\mathcal{G}_{m,q}(x_1, x_2)$ (see [23,24]; see also [25]), that is,

$$\mathcal{A}_{m,q}(x_1, x_2) = \mathcal{G}_{m,q}(x_1, x_2),$$

where $\mathcal{G}_{m,q}(x_1, x_2)$ are defined by

$$\frac{2t}{e_q(t) + 1} e_q(x_1 t) E_q(x_2 t) = \sum_{m=0}^{\infty} \mathcal{G}_{m,q}(x_1, x_2) \frac{t^m}{[m]_q!} \tag{18}$$

and $\mathcal{G}_{m,q} := \mathcal{G}_{m,q}(0,0)$ denotes the 2D q-Genocchi numbers.

We recall here that, in a recent paper, Khan and Riyasat [26] introduced the twice-iterated q-Appell polynomials $\mathcal{A}_{m,q}^{[2]}(x)$ which are defined by means of the following generating function:

$$\dot{A}_q(t) \ddot{A}_q(t) \, e_q(xt) = \sum_{m=0}^{\infty} \mathcal{A}_{m,q}^{[2]}(x) \frac{t^m}{[m]_q!} \quad (0 < q < 1). \tag{19}$$

In this paper, the class of the twice-iterated 2D q-Appell polynomials is introduced by means of generating functions, recurrence relations, partial q-difference equations, and series and determinant expressions. Further, several results are obtained for the members corresponding to this polynomial family. In Section 2, the twice-iterated 2D q-Appell polynomials are introduced by means of the generating functions and series definition. Also, the recurrence relation and q-difference equations involving the twice-iterated 2D q-Appell polynomials are derived. In Section 3, a determinant expression for the twice-iterated 2D q-Appell polynomials is established. In Section 4, the determinant expressions and some other properties of the members belonging to the family of the twice-iterated 2D q-Appell polynomials are obtained. Section 5 provides several graphical representations and surface plots associated with several members of families of q-polynomials which have investigated in this paper. Finally, in Section 6, we present some concluding remarks and observations.

2. Twice-Iterated 2D q-Appell Polynomials

In order to introduce the twice-iterated 2D q-Appell polynomials (2I2DqAP), we consider two different sets of the 2D q-Appell polynomials $\dot{A}_{m,q}(x_1, x_2)$ and $\ddot{A}_{m,q}(x_1, x_2)$ such that

$$\dot{A}_q(t)\, e_q(x_1 t) E_q(x_2 t) = \sum_{m=0}^{\infty} \dot{A}_{m,q}(x_1, x_2)\, \frac{t^m}{[m]_q!} \qquad (0 < q < 1), \tag{20}$$

where

$$\dot{A}_q(t) = \sum_{m=0}^{\infty} \dot{A}_{m,q}\, \frac{t^m}{[m]_q!}, \quad \dot{A}_q(t) \neq 0 \quad \text{and} \quad \dot{A}_{0,q} = 1; \tag{21}$$

$$\ddot{A}_q(t)\, e_q(x_1 t) E_q(x_2 t) = \sum_{m=0}^{\infty} \ddot{A}_{m,q}(x_1, x_2)\, \frac{t^m}{[m]_q!} \qquad (0 < q < 1), \tag{22}$$

where

$$\ddot{A}_q(t) = \sum_{m=0}^{\infty} \ddot{A}_{m,q}\, \frac{t^m}{[m]_q!}, \quad \ddot{A}_q(t) \neq 0 \quad \text{and} \quad \ddot{A}_{0,q} = 1; \tag{23}$$

$$\ddot{A}_q(t)\, e_q(x_1 t) = \sum_{m=0}^{\infty} \ddot{A}_{m,q}(x_1)\, \frac{t^m}{[m]_q!} \qquad (0 < q < 1). \tag{24}$$

The generating function for the 2I2DqAP is asserted by the following result.

Theorem 1. *The generating function for the twice-iterated 2D q-Appell polynomials $A_{m,q}^{[2]}(x_1, x_2)$ is given by*

$$\dot{A}_q(t)\ddot{A}_q(t)\, e_q(x_1 t) E_q(x_2 t) = \sum_{m=0}^{\infty} A_{m,q}^{[2]}(x_1, x_2)\, \frac{t^m}{[m]_q!} \qquad (0 < q < 1). \tag{25}$$

Proof. By expanding the first q-exponential function $e_q(x_1 t)$ in the left-hand side of the Equation (20) and then replacing the powers of x, that is, $x_1^0, x_1, x_1^2, \cdots, x_1^m$ by the polynomials $\ddot{A}_{0,q}(x_1), \ddot{A}_{1,q}(x_1), \ddot{A}_{2,q}(x_1), \cdots, \ddot{A}_{m,q}(x_1)$ in the left-hand side and x_1 by $\ddot{A}_{1,q}(x_1)$ in the right-hand side of the resultant equation, we have

$$\dot{A}_q(t)\left(1 + \ddot{A}_{1,q}(x_1)\, \frac{t}{[1]_q!} + \ddot{A}_{2,q}(x_1)\, \frac{t^2}{[2]_q!} + \cdots + \ddot{A}_{m,q}(x_1)\, \frac{t^m}{[m]_q!} + \cdots\right) E_q(x_2 t) = \sum_{m=0}^{\infty} \dot{A}_{m,q}(\ddot{A}_{1,q}(x_1), x_2)\, \frac{t^m}{[m]_q!}. \tag{26}$$

Moreover, by summing up the series in the left-hand side and then using the Equation (24) in the resulting equation, we get

$$\dot{A}_q(t)\, \ddot{A}_q(t)\, e_q(x_1 t) E_q(x_2 t) = \sum_{m=0}^{\infty} \dot{A}_{m,q}(\ddot{A}_{1,q}(x_1), x_2)\, \frac{t^m}{[m]_q!}. \tag{27}$$

Finally, denoting the resulting 2I2DqAP in the right-hand side of the above equation by $A_{m,q}^{[2]}(x_1, x_2)$, that is,

$$\dot{A}_{m,q}(\ddot{A}_{1,q}(x_1), x_2) = A_{m,q}^{[2]}(x_1, x_2), \tag{28}$$

the assertion (25) of Theorem 1 is proved. $\square$

Remark 1. *For* $x_2 = 0$, *the 2I2DqAP* $\mathcal{A}_{m,q}^{[2]}(x_1, x_2)$ *reduce to the twice-iterated q-Appell polynomials (see* [26]*) such that*

$$\mathcal{A}_{m,q}^{[2]}(x_1) := \mathcal{A}_{m,q}^{[2]}(x_1, 0). \tag{29}$$

It is also noted that

$$\mathcal{A}_{m,q} := \mathcal{A}_{m,q}(0) = \mathcal{A}_{m,q}(0,0). \tag{30}$$

We next give the series definition for the 2I2DqAP $\mathcal{A}_{m,q}^{[2]}(x_1, x_2)$ by proving the following result.

Theorem 2. *The twice-iterated 2D q-Appell polynomials* $\mathcal{A}_{m,q}^{[2]}(x_1, x_2)$ *are given by the following series expression:*

$$\mathcal{A}_{m,q}^{[2]}(x_1, x_2) = \sum_{s=0}^{m} \begin{bmatrix} m \\ s \end{bmatrix}_q \dot{\mathcal{A}}_{s,q} \ddot{\mathcal{A}}_{m-s,q}(x_1, x_2). \tag{31}$$

Proof. In view of the Equations (21) and (22), the Equation (25) can be written as follows:

$$\sum_{s=0}^{\infty} \dot{\mathcal{A}}_{s,q} \frac{t^s}{[s]_q!} \sum_{m=0}^{\infty} \ddot{\mathcal{A}}_{m,q}(x_1, x_2) \frac{t^m}{[m]_q!} = \sum_{m=0}^{\infty} \mathcal{A}_{m,q}^{[2]}(x_1, x_2) \frac{t^m}{[m]_q!}, \tag{32}$$

which, on using the Cauchy product rule, gives

$$\sum_{m=0}^{\infty} \sum_{s=0}^{m} \begin{bmatrix} m \\ s \end{bmatrix}_q \dot{\mathcal{A}}_{s,q} \ddot{\mathcal{A}}_{m-s,q}(x_1, x_2) \frac{t^m}{[m]_q!} = \sum_{m=0}^{\infty} \mathcal{A}_{m,q}^{[2]}(x_1, x_2) \frac{t^m}{[m]_q!}. \tag{33}$$

Equating the coefficients of like powers of t in both sides of the above equation, we arrive at the assertion (31) of Theorem 2. $\square$

Remark 2. For $x_2 = 0$, the series expression (31) becomes

$$\mathcal{A}_{m,q}^{[2]}(x_1) = \sum_{s=0}^{m} \begin{bmatrix} m \\ s \end{bmatrix}_q \dot{\mathcal{A}}_{s,q} \ddot{\mathcal{A}}_{m-s,q}(x_1), \tag{34}$$

which is a known result [26] (p. 5, Equation (2.8)).

We now state and prove the following result.

Theorem 3. *The following relation between the twice-iterated 2D q-Appell polynomials* $\mathcal{A}_{m,q}^{[2]}(x_1, x_2)$ *and the twice-iterated q-Appell polynomials* $\mathcal{A}_{m,q}(x_1)$ *holds true:*

$$\mathcal{A}_{m,q}^{[2]}(x_1, x_2) = \sum_{s=0}^{m} \begin{bmatrix} m \\ s \end{bmatrix}_q q^{\frac{1}{2}s(s-1)} x_2^s \, \mathcal{A}_{m-s,q}^{[2]}(x_1). \tag{35}$$

Proof. Using the Equations (7) and (19) in the left-hand side of the generating function (25), we get

$$\sum_{m=0}^{\infty} \mathcal{A}_{m,q}^{[2]}(x_1, x_2) \frac{t^m}{[m]_q!} = \left(\sum_{m=0}^{\infty} \mathcal{A}_{m,q}^{[2]}(x_1) \frac{t^m}{[m]_q!} \right) \left(\sum_{m=0}^{\infty} q^{\frac{1}{2}m(m-1)} \frac{(x_2 t)^m}{[m]_q!} \right), \tag{36}$$

which, on applying the Cauchy product rule in the left-hand side, yields

$$\sum_{m=0}^{\infty} \mathcal{A}_{m,q}^{[2]}(x_1,x_2)\,\frac{t^m}{[m]_q!} = \sum_{m=0}^{\infty}\sum_{s=0}^{m} \begin{bmatrix} m \\ s \end{bmatrix}_q q^{\frac{1}{2}s(s-1)} x_2^s \, \mathcal{A}_{m-s,q}^{[2]}(x_1)\,\frac{t^m}{[m]_q!}. \tag{37}$$

Finally, equating the coefficients of like powers of t on both sides of this last equation, we obtain the assertion (35) of Theorem 3. $\square$

Remark 3. *By taking* $x_2 = 1$ *in the result* (35), *we get*

$$\mathcal{A}_{m,q}^{[2]}(x_1,1) = \sum_{s=0}^{m} \begin{bmatrix} m \\ s \end{bmatrix}_q q^{\frac{1}{2}s(s-1)} \, \mathcal{A}_{m-s,q}^{[2]}(x_1). \tag{38}$$

Remark 4. *The following statements are equivalent:*

$$(a) \quad \mathcal{A}_{m,q}^{[2]}(x_1,-x_2) = (-1)^m \mathcal{A}_{m,q}^{[2]}(0,x_2) \tag{39}$$

and

$$(b) \quad \mathcal{A}_{m,q}^{[2]}(x_1) = (-1)^m \mathcal{A}_{m,q}^{[2]}(0) \tag{40}$$

In order to derive the q-recurrence relations and the q-difference equations for the twice-iterated 2D q-Appell polynomials by using the lowering operators that are, in fact, the q-derivative operator D_q, we first prove the following lemma.

Lemma 1. *The twice-iterated 2D q-Appell polynomials* $\mathcal{A}_{m,q}^{[2]}(x_1,x_2)$ *satisfy the following operational relations:*

$$D_{q,x_1}\left(\mathcal{A}_{m,q}^{[2]}(x_1,x_2)\right) = [m]_q \, \mathcal{A}_{m-1,q}^{[2]}(x_1,x_2), \tag{41}$$

$$D_{q,x_2}\left(\mathcal{A}_{m,q}^{[2]}(x_1,x_2)\right) = [m]_q \, \mathcal{A}_{m-1,q}^{[2]}(x_1,qx_2), \tag{42}$$

$$\mathcal{A}_{m-s,q}^{[2]}(x_1,x_2) = \frac{[m-s]_q!}{[m]_q!} \, D_{q,x_1}^s \, \mathcal{A}_{m,q}^{[2]}(x_1,x_2) \tag{43}$$

and

$$q^{\frac{s(s-1)}{2}} \, \mathcal{A}_{m-s,q}^{[2]}(x_1,q^s x_2) = \frac{[m-s]_q!}{[m]_q!} D_{q,x_2}^s \mathcal{A}_{m,q}^{[2]}(x_1,x_2). \tag{44}$$

Proof. In view of the Equation (25), the proof of the above lemma requires a direct use of the identity (5). We, therefore, skip the details involved. $\square$

We now derive the q-recurrence relations for the 2I2DqAP $\mathcal{A}_{m,q}^{[2]}(x_1,x_2)$.

Theorem 4. *The twice-iterated 2D q-Appell polynomials* $\mathcal{A}_{m,q}^{[2]}(x_1,x_2)$ *satisfy the following linear homogeneous recurrence relation:*

$$\mathcal{A}_{m,q}^{[2]}(qx_1,x_2) = \frac{1}{[m]_q} \sum_{s=0}^{m} \begin{bmatrix} m \\ s \end{bmatrix}_q q^{m-s}(\alpha_s + x_2\beta_s + \gamma_s)\mathcal{A}_{m-s,q}^{[2]}(x_1,x_2) + x_1 q^m \, \mathcal{A}_{m-1,q}^{[2]}(x_1,x_2), \tag{45}$$

where

$$t\frac{\ddot{A}_q(t)D_{q,t}A_q(t)}{A_q(qt)\ddot{A}_q(qt)} = \sum_{m=0}^{\infty} \alpha_m \frac{t^m}{[m]_q!}, \qquad t\frac{A_q(t)\ddot{A}_q(t)}{A_q(qt)\ddot{A}_q(qt)} = \sum_{m=0}^{\infty} \beta_m \frac{t^m}{[m]_q!},$$

$$t\frac{D_{q,t}\ddot{A}_q(t)}{\ddot{A}_q(qt)} = \sum_{m=0}^{\infty} \gamma_m \frac{t^m}{[m]_q!}. \tag{46}$$

Proof. Consider the following generating function:

$$G_q(qx_1, x_2, t) = \dot{A}_q(t)\ddot{A}_q(t)\, e_q(qx_1 t)E_q(x_2 t) = \sum_{m=0}^{\infty} A_{m,q}^{[2]}(qx_1, x_2)\frac{t^m}{[m]_q!}. \tag{47}$$

By taking the q-derivative of the Equation (47) partially with respect to t, we get

$$D_{q,t}\big(G_q(qx_1, x_2, t)\big) = x_2 A_q(t)\ddot{A}_q(t)\, e_q(qxt)\, E_q(qx_2 t) + qx_1 A_q(qt)\,\ddot{A}_q(qt)\, e_q(qxt)\, E_q(qx_2 t)$$
$$+ (D_{q,t}A_q(t))\ddot{A}_q(t)e_q(qxt)\, E_q(qx_2 t) + A_q(qt)(D_{q,t}\ddot{A}_q(t))\, e_q(qxt)\, E_q(qx_2 t). \tag{48}$$

Thus, upon factorizing $G_q(qx_1, x_2, t)$ occurring in the left-hand side and multiplying both sides of the identity (48) by t, we find that

$$tD_{q,t}\big(G_q(qx_1, x_2, t)\big)$$
$$= G_q(qx_1, x_2, t)\left(t\frac{\ddot{A}_q(t)D_{q,t}A_q(t)}{A_t(qt)\ddot{A}_q(qt)} + x_2 t\frac{A_q(t)\ddot{A}_q(t)}{A_q(qt)\ddot{A}_q(qt)} + t\frac{D_{q,t}\ddot{A}_q(t)}{\ddot{A}_q(qt)} + qtx_1\right). \tag{49}$$

In view of the assumption (46) and the Equation (47), the Equation (49) becomes

$$\sum_{m=0}^{\infty} [m]_q\, A_{m,q}^{[2]}(qx_1, x_2)\,\frac{t^m}{[m]_q!}$$
$$= \sum_{m=0}^{\infty} q^m A_{m,q}^{[2]}(x_1, x_2)\frac{t^m}{[m]_q!}\left(\sum_{m=0}^{\infty} \alpha_m \frac{t^m}{[m]_q!} + x_2 \sum_{m=0}^{\infty} \beta_m \frac{t^m}{[m]_q!} + \sum_{m=0}^{\infty} \gamma_m \frac{t^m}{[m]_q!} + qx_1\right), \tag{50}$$

which, on using the Cauchy product rule, gives

$$\sum_{m=0}^{\infty} [m]_q\, A_{m,q}^{[2]}(qx_1, x_2)\,\frac{t^m}{[m]_q!}$$
$$= \sum_{m=0}^{\infty}\sum_{s=0}^{m} \begin{bmatrix} m \\ s \end{bmatrix}_q q^{m-s}(\alpha_s + x_2\beta_s + \gamma_s)A_{m-s,q}^{[2]}(x_1, x_2)\frac{t^m}{[m]_q!} \tag{51}$$
$$+ x_1 \sum_{m=0}^{\infty} [m]_q q^m A_{m-1,q}^{[2]}(x_1, x_2)\,\frac{t^m}{[m]_q!}.$$

Finally, upon equating the coefficients of like powers of t on both sides of the above equation and dividing both sides of the resulting equation by $[m]_q$, we get the assertion (45) of Theorem 4. $\square$

We now state and prove the following result.

Theorem 5. *The following recurrence relation for the twice-iterated* 2D *q-Appell polynomials* $\mathcal{A}_{m,q}^{[2]}(x_1,x_2)$ *holds true:*

$$\mathcal{A}_{m,q}^{[2]}(qx_1,x_2) = q^{m-1}\left(\frac{\ddot{A}_q(t)D_{q,t}\dot{A}_q(t)}{\dot{A}_q(qt)\ddot{A}_q(qt)} + x_2\frac{\dot{A}_q(t)\ddot{A}_q(t)}{\dot{A}_q(qt)\,\ddot{A}_q(qt)} + \frac{D_{q,t}\ddot{A}_q(t)}{\ddot{A}_q(qt)} + qx_1\right)\mathcal{A}_{m-1,q}^{[2]}(x_1,x_2). \quad (52)$$

Proof. We first use the Equation (47) in both sides of the Equation (49). Then, after some simplification, by equating the coefficients of like powers of t on both sides of the resulting equation, we arrive at the assertion (52) of Theorem 5. $\square$

We next derive the q-difference equations which are satisfied by the twice-iterated 2D q-Appell polynomials.

Theorem 6. *The twice-iterated* 2D *q-Appell polynomials* $\mathcal{A}_{m,q}^{[2]}(x_1,x_2)$ *are the solutions of the following* q-*difference equations:*

$$\left(\sum_{s=0}^{m}\frac{q^{m-s}}{[s]_q}(\alpha_s + x_2\beta_s + \gamma_s)D_{q,x_1}^s + x_1 q^m D_{q,x_1}\right)\mathcal{A}_{m,q}^{[2]}(x_1,x_2) - [m]_q\mathcal{A}_{m,q}^{[2]}(qx_1,x_2) = 0 \quad (53)$$

or

$$\sum_{s=0}^{m}\frac{q^{m-s}}{[s]_q}\left(\alpha_s + x_2\frac{\beta_s}{q^s} + \gamma_s\right)D_{q,x_2}^s\mathcal{A}_{m,q}^{[2]}\left(x_1,\frac{x_2}{q^s}\right) + x_1 q^m D_{q,x_2}\mathcal{A}_{m,q}^{[2]}\left(x_1,\frac{x_2}{q}\right) - [m]_q\mathcal{A}_{m,q}^{[2]}(qx_1,x_2) = 0. \quad (54)$$

Proof. The proof of the assertions (53) and (54) of Theorem 6 would follow directly upon using the Equations (43) and (44), respectively, in the recurrence relation (45). $\square$

In the next section (Section 3 below), the determinant forms for the 2I2DqAP are established.

3. The Twice-Iterated 2D q-Appell Polynomials from the Determinant Viewpoint

One of the important aspects of the study of any polynomial system is to find its potentially useful determinant representation. Recently, Keleshteri and Mahmudov [21] introduced the determinant definitions for the q-Appell polynomials and the 2D q-Appell polynomials. These polynomials are useful in finding the solutions of some general linear interpolation problems and can also be used for computational purposes. Khan and Riyasat [26], on the other hand, established the determinant expressions for the twice-iterated q-Appell polynomials. This fact provides motivation for us to establish the determinant definitions and the determinant expressions for the twice-iterated 2D q-Appell polynomials 2I2DqAP by proving the following result.

Theorem 7. *The 2I2DqAP $A_{m,q}^{[2]}(x_1, x_2)$ of degree m are defined by*

$$A_{0,q}^{[2]}(x_1, x_2) = \frac{1}{\mathcal{B}_{0,q}}, \tag{55}$$

$$A_{m,q}^{[2]}(x_1, x_2) = \frac{(-1)^m}{(\mathcal{B}_{0,q})^{m+1}} \begin{vmatrix} 1 & \ddot{A}_{1,q}(x_1, x_2) & \ddot{A}_{2,q}(x_1, x_2) & \cdots & \ddot{A}_{m-1,q}(x_1, x_2) & \ddot{A}_m(x_1, x_2) \\ \mathcal{B}_{0,q} & \mathcal{B}_{1,q} & \mathcal{B}_{2,q} & \cdots & \mathcal{B}_{m-1,q} & \mathcal{B}_{m,q} \\ 0 & \mathcal{B}_{0,q} & \begin{bmatrix}2\\1\end{bmatrix}_q \mathcal{B}_{1,q} & \cdots & \begin{bmatrix}m-1\\1\end{bmatrix}_q \mathcal{B}_{m-2,q} & \begin{bmatrix}m\\1\end{bmatrix}_q \mathcal{B}_{m-1,q} \\ 0 & 0 & \mathcal{B}_{0,q} & \cdots & \begin{bmatrix}m-1\\2\end{bmatrix}_q \mathcal{B}_{m-3,q} & \begin{bmatrix}m\\2\end{bmatrix}_q \mathcal{B}_{m-2,q} \\ \vdots & \vdots & \vdots & \ddots & \vdots & \vdots \\ 0 & 0 & 0 & \cdots & \mathcal{B}_{0,q} & \begin{bmatrix}m\\m-1\end{bmatrix}_q \mathcal{B}_{1,q} \end{vmatrix}, \tag{56}$$

$$\mathcal{B}_{m,q} = -\frac{1}{\ddot{A}_{0,q}} \left(\sum_{s=1}^{m} \begin{bmatrix}m\\s\end{bmatrix}_q \ddot{A}_{s,q} \mathcal{B}_{m-s,q} \right) \qquad (m \in \mathbb{N}),$$

where $\mathcal{B}_{0,q} \neq 0$, $\mathcal{B}_{0,q} = \frac{1}{\ddot{A}_{0,q}}$ and $\ddot{A}_{m,q}(x_1, x_2)$ $(m \in \mathbb{N}_0)$ are the q-Appell polynomials of degree m.

Proof. Consider $A_{m,q}^{[2]}(x_1, x_2)$ as a sequence of the 2I2DqAP defined by the Equation (25). Also let $\ddot{A}_{m,q}$ and $\mathcal{B}_{m,q}$ be two numerical sequences (the coefficients of the q-Taylor series expansions of functions) such that

$$\ddot{A}_q(t) = \ddot{A}_{0,q} + \ddot{A}_{1,q} \frac{t}{[1]_q!} + \ddot{A}_{2,q} \frac{t^2}{[2]_q!} + \cdots + \ddot{A}_{m,q} \frac{t^m}{[m]_q!} + \cdots \qquad (m \in \mathbb{N}_0;\ \ddot{A}_{0,q} \neq 0) \tag{57}$$

and

$$\hat{\ddot{A}}_q(t) = \mathcal{B}_{0,q} + \mathcal{B}_{1,q} \frac{t}{[1]_q!} + \mathcal{B}_{2,q} \frac{t^2}{[2]_q!} + \cdots + \mathcal{B}_{m,q} \frac{t^m}{[m]_q!} + \cdots \qquad (m \in \mathbb{N}_0;\ \mathcal{B}_{0,q} \neq 0), \tag{58}$$

also satisfying the following condition:

$$\ddot{A}_q(t) \hat{\ddot{A}}_q(t) = 1. \tag{59}$$

On using the Cauchy product rule for the two-series product $\ddot{A}_q(t) \hat{\ddot{A}}_q(t)$, we get

$$\ddot{A}_q(t) \hat{\ddot{A}}_q(t) = \sum_{m=0}^{\infty} \ddot{A}_{m,q} \frac{t^m}{[m]_q!} \sum_{m=0}^{\infty} \mathcal{B}_{m,q} \frac{t^m}{[m]_q!}$$

$$= \sum_{m=0}^{\infty} \sum_{s=0}^{m} \begin{bmatrix}m\\s\end{bmatrix}_q \ddot{A}_{s,q} \mathcal{B}_{m-s,q} \frac{t^m}{[m]_q!}.$$

Consequently, we have

$$\sum_{s=0}^{m} \begin{bmatrix}m\\s\end{bmatrix}_q \ddot{A}_{s,q} \mathcal{B}_{m-s,q} = \begin{cases} 1, & if\ m = 0, \\ 0, & if\ m \in \mathbb{N}. \end{cases} \tag{60}$$

that is,

$$\begin{cases} \mathcal{B}_{0,q} = \frac{1}{\ddot{A}_{0,q}}; \\[2mm] \mathcal{B}_{m,q} = -\frac{1}{\ddot{A}_{0,q}} \left(\sum_{s=1}^{m} \begin{bmatrix}m\\s\end{bmatrix}_q \ddot{A}_{s,q} \mathcal{B}_{m-s,q} \right) \qquad (m \in \mathbb{N}_0). \end{cases} \tag{61}$$

Next, upon multiplying both sides of the Equation (25) by $\hat{\ddot{A}}_q(t)$, we get

$$\ddot{A}_q(t) \hat{\ddot{A}}_q(t)\, \ddot{A}_q(t)\, e_q(x_1 t) E_q(x_2 t) = \hat{\ddot{A}}_q(t) \sum_{m=0}^{\infty} A_{m,q}^{[2]}(x_1, x_2) \frac{t^m}{[m]_q!}. \tag{62}$$

Further, in view of the Equations (22), (58) and (59), the above Equation (62) becomes

$$\sum_{m=0}^{\infty} \ddot{A}_{m,q}(x_1,x_2)\, \frac{t^m}{[m]_q!} = \sum_{m=0}^{\infty} \mathcal{B}_{m,q}\, \frac{t^m}{[m]_q!} \sum_{m=0}^{\infty} A_{m,q}^{[2]}(x_1,x_2)\, \frac{t^m}{[m]_q!}. \tag{63}$$

Now, on using the Cauchy product rule for the two series in the right-hand side of the Equation (63), we obtain the following infinite system for the unknowns $A_{m,q}^{[2]}(x_1,x_2)$:

$$\begin{cases}
\mathcal{B}_{0,q}\, A_{0,q}^{[2]}(x_1,x_2) = 1; \\[2ex]
\mathcal{B}_{1,q}\, A_{0,q}^{[2]}(x_1,x_2) + \mathcal{B}_{0,q}\, A_{1,q}^{[2]}(x_1,x_2) = \ddot{A}_{1,q}(x_1,x_2), \\[2ex]
\mathcal{B}_{2,q}\, A_{0,q}^{[2]}(x_1,x_2) + \left[{}^2_1\right]_q \mathcal{B}_{1,q}\, A_{1,q}^{[2]}(x_1,x_2) + \mathcal{B}_{0,q}\, A_{2,q}^{[2]}(x_1,x_2) = \ddot{A}_{2,q}(x_1,x_2), \\[1ex]
\;\;\vdots \\[1ex]
\mathcal{B}_{m-1,q}\, A_{0,q}^{[2]}(x_1,x_2) + \left[{}^{m-1}_{\;1}\right]_q \mathcal{B}_{m-2,q}\, A_{1,q}^{[2]}(x_1,x_2) + \cdots + \mathcal{B}_{0,q}\, A_{m-1,q}^{[2]}(x_1,x_2) = \ddot{A}_{m-1,q}(x_1,x_2), \\[2ex]
\mathcal{B}_{m,q}\, A_{0,q}^{[2]}(x_1,x_2) + \left[{}^m_1\right]_q \mathcal{B}_{m-1,q}\, A_{1,q}^{[2]}(x_1,x_2) + \cdots + \mathcal{B}_{0,q}\, A_{m,q}^{[2]}(x_1,x_2) = \ddot{A}_{m,q}(x_1,x_2), \\[1ex]
\;\;\vdots
\end{cases} \tag{64}$$

Obviously, the first equation of the system (64) leads to our first assertion (55). The coefficient matrix of the system (64) is lower triangular, so this helps us to obtain the unknowns $A_{m,q}^{[2]}(x_1,x_2)$ by applying the Cramer rule to the first $m+1$ equations of the system (64). Accordingly, we can obtain

$$A_{m,q}^{[2]}(x_1,x_2) = \frac{\begin{vmatrix}
\mathcal{B}_{0,q} & 0 & 0 & \cdots & 0 & 1 \\
\mathcal{B}_{1,q} & \mathcal{B}_{0,q} & 0 & \cdots & 0 & \ddot{A}_{1,q}(x_1,x_2) \\
\mathcal{B}_{2,q} & \left[{}^2_1\right]_q \mathcal{B}_{1,q} & \mathcal{B}_{0,q} & \cdots & 0 & \ddot{A}_{2,q}(x_1,x_2) \\
\vdots & \vdots & \vdots & \ddots & \vdots & \vdots \\
\mathcal{B}_{m-1,q} & \left[{}^{m-1}_{\;1}\right]_q \mathcal{B}_{m-2,q} & \left[{}^{m-1}_{\;2}\right]_q \mathcal{B}_{m-3,q} & \cdots & \mathcal{B}_{0,q} & \ddot{A}_{m-1,q}(x_1,x_2) \\
\mathcal{B}_{m,q} & \left[{}^m_1\right]_q \mathcal{B}_{m-1,q} & \left[{}^m_2\right]_q \mathcal{B}_{m-2,q} & \cdots & \left[{}^{\;m}_{m-1}\right]_q \mathcal{B}_{1,q} & \ddot{A}_{m,q}(x_1,x_2)
\end{vmatrix}}{\begin{vmatrix}
\mathcal{B}_{0,q} & 0 & 0 & \cdots & 0 & 1 \\
\mathcal{B}_{1,q} & \mathcal{B}_{0,q} & 0 & \cdots & 0 & 0 \\
\mathcal{B}_{2,q} & \left[{}^2_1\right]_q \mathcal{B}_{1,q} & \mathcal{B}_{0,q} & \cdots & 0 & 0 \\
\vdots & \vdots & \vdots & \ddots & \vdots & \vdots \\
\mathcal{B}_{m-1,q} & \left[{}^{m-1}_{\;1}\right]_q \mathcal{B}_{m-2,q} & \left[{}^{m-1}_{\;2}\right]_q \mathcal{B}_{m-3,q} & \cdots & \mathcal{B}_{0,q} & 0 \\
\mathcal{B}_{m,q} & \left[{}^m_1\right]_q \mathcal{B}_{m-1,q} & \left[{}^m_2\right]_q \mathcal{B}_{m-2,q} & \cdots & \left[{}^{\;m}_{m-1}\right]_q \mathcal{B}_{1,q} & \mathcal{B}_{0,q}
\end{vmatrix}}, \tag{65}$$

where $m \in \mathbb{N}$. Thus, upon expanding the determinant in the denominator and taking the transpose of the determinant in the numerator, we get

$$
A_{m,q}^{[2]}(x_1, x_2) = \frac{1}{(\mathcal{B}_{0,q})^{m+1}}
\begin{vmatrix}
\mathcal{B}_{0,q} & \mathcal{B}_{1,q} & \mathcal{B}_{2,q} & \cdots & \mathcal{B}_{m-1,q} & \mathcal{B}_{n,q} \\
0 & \mathcal{B}_{0,q} & \left[\begin{smallmatrix}2\\1\end{smallmatrix}\right]_q \mathcal{B}_{1,q} & \cdots & \left[\begin{smallmatrix}m-1\\1\end{smallmatrix}\right]_q \mathcal{B}_{m-2,q} & \left[\begin{smallmatrix}m\\1\end{smallmatrix}\right]_q \mathcal{B}_{m-1,q} \\
0 & 0 & \mathcal{B}_{0,q} & \cdots & \left[\begin{smallmatrix}m-1\\2\end{smallmatrix}\right]_q \mathcal{B}_{m-3,q} & \left[\begin{smallmatrix}m\\2\end{smallmatrix}\right]_q \mathcal{B}_{m-2,q} \\
\vdots & \vdots & \vdots & \ddots & \vdots & \vdots \\
0 & 0 & 0 & \cdots & \mathcal{B}_{0,q} & \left[\begin{smallmatrix}m\\m-1\end{smallmatrix}\right]_q \mathcal{B}_{1,q} \\
1 & \ddot{A}_{1,q}(x_1, x_2) & \ddot{A}_{2,q}(x_1, x_2) & \cdots & \ddot{A}_{m-1,q}(x_1, x_2) & \ddot{A}_{m,q}(x_1, x_2)
\end{vmatrix}.
\tag{66}
$$

Finally, after m circular row exchanges, that is, after moving the jth row to the $(j+1)$st position for $j = 1, 2, 3, \cdots, m-1$, we arrive at our assertion (56) of Theorem 7. $\square$

Theorem 8. *The following identity for the 2I2DqAP $A_{m,q}^{[2]}(x_1, x_2)$ holds true:*

$$
A_{m,q}^{[2]}(x_1, x_2) = \frac{1}{\mathcal{B}_{0,q}} \left(\ddot{A}_{m,q}(x_1, x_2) - \sum_{s=0}^{m-1} \begin{bmatrix} m \\ s \end{bmatrix}_q \mathcal{B}_{m-s,q} \, A_{s,q}^{[2]}(x_1, x_2) \right) \quad (m \in \mathbb{N}).
\tag{67}
$$

Proof. Expanding the determinant in the Equation (56) with respect to the $(m+1)^{\text{st}}$ row, we get

$$
A_{m,q}^{[2]}(x_1, x_2) = \frac{(-1)^m}{(\mathcal{B}_{0,q})^{m+1}} \begin{bmatrix} m \\ m-1 \end{bmatrix}_q \mathcal{B}_{1,q}
\begin{vmatrix}
1 & \ddot{A}_{1,q}(x_1, x_2) & \ddot{A}_{2,q}(x_1, x_2) & \cdots & \cdots & \ddot{A}_{m-1,q}(x_1, x_2) \\
\mathcal{B}_{0,q} & \mathcal{B}_{1,q} & \mathcal{B}_{2,q} & \cdots & \cdots & \mathcal{B}_{m-1,q} \\
0 & \mathcal{B}_{0,q} & \left[\begin{smallmatrix}2\\1\end{smallmatrix}\right]_q \mathcal{B}_{1,q} & \cdots & \cdots & \left[\begin{smallmatrix}m-1\\1\end{smallmatrix}\right]_q \mathcal{B}_{m-2,q} \\
0 & 0 & \mathcal{B}_{0,q} & \cdots & \cdots & \left[\begin{smallmatrix}m-1\\2\end{smallmatrix}\right]_q \mathcal{B}_{m-3,q} \\
\vdots & \vdots & \vdots & \ddots & \vdots & \vdots \\
0 & 0 & 0 & \cdots & \mathcal{B}_{0,q} & \left[\begin{smallmatrix}m-1\\m-2\end{smallmatrix}\right]_q \mathcal{B}_{1,q}
\end{vmatrix}
$$

$$
+ \frac{(-1)^m}{(\mathcal{B}_{0,q})^{m+1}}
\begin{vmatrix}
1 & \ddot{A}_{1,q}(x_1, x_2) & \ddot{A}_{2,q}(x_1, x_2) & \cdots & \cdots & \ddot{A}_{m-2,q}(x_1, x_2) & \ddot{A}_{m,q}(x_1, x_2) \\
\mathcal{B}_{0,q} & \mathcal{B}_{1,q} & \mathcal{B}_{2,q} & \cdots & \cdots & \mathcal{B}_{m-1,q} & \mathcal{B}_{m-1,q} \\
0 & \mathcal{B}_{0,q} & \left[\begin{smallmatrix}2\\1\end{smallmatrix}\right]_q \mathcal{B}_{1,q} & \cdots & \cdots & \left[\begin{smallmatrix}m-2\\1\end{smallmatrix}\right]_q \mathcal{B}_{m-3,q} & \left[\begin{smallmatrix}m-1\\1\end{smallmatrix}\right]_q \mathcal{B}_{m-2,q} \\
0 & 0 & \mathcal{B}_{0,q} & \cdots & \cdots & \left[\begin{smallmatrix}m-2\\2\end{smallmatrix}\right]_q \mathcal{B}_{m-4,q} & \left[\begin{smallmatrix}m-1\\2\end{smallmatrix}\right]_q \mathcal{B}_{m-3,q} \\
\vdots & \vdots & \vdots & \ddots & \vdots & \vdots & \vdots \\
0 & 0 & 0 & \cdots & \cdots & \mathcal{B}_{0,q} & \left[\begin{smallmatrix}m-1\\m-2\end{smallmatrix}\right]_q \mathcal{B}_{1,q}
\end{vmatrix}
$$

$$
= \frac{-1}{\mathcal{B}_{0,q}} \begin{bmatrix} m \\ m-1 \end{bmatrix}_q \mathcal{B}_{1,q} \mathcal{A}^{[2]}_{m-1,q}(x_1, x_2) + \frac{(-1)^m}{(\mathcal{B}_{0,q})^m}
$$

$$
\begin{vmatrix}
1 & \ddot{\mathcal{A}}_{1,q}(x_1,x_2) & \ddot{\mathcal{A}}_{2,q}(x_1,x_2) & \cdots & \cdots & \ddot{\mathcal{A}}_{m-2,q}(x_1,x_2) & \ddot{\mathcal{A}}_{m,q}(x_1,x_2) \\
\mathcal{B}_{0,q} & \mathcal{B}_{1,q} & \mathcal{B}_{2,q} & \cdots & \cdots & \mathcal{B}_{m-1,q} & \mathcal{B}_{m-1,q} \\
0 & \mathcal{B}_{0,q} & \begin{bmatrix}2\\1\end{bmatrix}_q \mathcal{B}_{1,q} & \cdots & \cdots & \begin{bmatrix}m-2\\1\end{bmatrix}_q \mathcal{B}_{m-3,q} & \begin{bmatrix}m-1\\1\end{bmatrix}_q \mathcal{B}_{m-2,q} \\
0 & 0 & \mathcal{B}_{0,q} & \cdots & \cdots & \begin{bmatrix}m-2\\2\end{bmatrix}_q \mathcal{B}_{m-4,q} & \begin{bmatrix}m-1\\2\end{bmatrix}_q \mathcal{B}_{m-3,q} \\
\vdots & \vdots & \vdots & \ddots & \vdots & \vdots & \vdots \\
0 & 0 & 0 & \cdots & \cdots & \mathcal{B}_{0,q} & \begin{bmatrix}m-1\\m-2\end{bmatrix}_q \mathcal{B}_{1,q}
\end{vmatrix} .
$$

Next, by applying the same argument for the last determinant, we find that

$$
\mathcal{A}^{[2]}_{m,q}(x_1, x_2) = \frac{-1}{\mathcal{B}_{0,q}} \begin{bmatrix} m \\ m-1 \end{bmatrix}_q \mathcal{B}_{1,q} \mathcal{A}^{[2]}_{m-1,q}(x_1, x_2) + \frac{(-1)^m}{(\mathcal{B}_{0,q})^m} \begin{bmatrix} m-1 \\ m-2 \end{bmatrix}_q \mathcal{B}_{2,q}
$$

$$
\begin{vmatrix}
1 & \ddot{\mathcal{A}}_{1,q}(x_1,x_2) & \ddot{\mathcal{A}}_{2,q}(x_1,x_2) & \cdots & \cdots & \ddot{\mathcal{A}}_{m-3,q}(x_1,x_2) & \ddot{\mathcal{A}}_{m-2,q}(x_1,x_2) \\
\mathcal{B}_{0,q} & \mathcal{B}_{1,q} & \mathcal{B}_{2,q} & \cdots & \cdots & \mathcal{B}_{m-3,q} & \mathcal{B}_{m-2,q} \\
0 & \mathcal{B}_{0,q} & \begin{bmatrix}2\\1\end{bmatrix}_q \mathcal{B}_{1,q} & \cdots & \cdots & \begin{bmatrix}m-3\\1\end{bmatrix}_q \mathcal{B}_{m-4,q} & \begin{bmatrix}m-2\\1\end{bmatrix}_q \mathcal{B}_{m-3,q} \\
0 & 0 & \mathcal{B}_{0,q} & \cdots & \cdots & \begin{bmatrix}m-3\\2\end{bmatrix}_q \mathcal{B}_{m-5,q} & \begin{bmatrix}m-2\\2\end{bmatrix}_q \mathcal{B}_{m-4,q} \\
\vdots & \vdots & \vdots & \ddots & \vdots & \vdots & \vdots \\
0 & 0 & 0 & \cdots & \cdots & \mathcal{B}_{0,q} & \begin{bmatrix}m-2\\m-3\end{bmatrix}_q \mathcal{B}_{1,q}
\end{vmatrix}
$$

$$
+ \frac{(-1)^{m+2}}{(\mathcal{B}_{0,q})^m} \mathcal{B}_{0,q}
\begin{vmatrix}
1 & \ddot{\mathcal{A}}_{1,q}(x_1,x_2) & \ddot{\mathcal{A}}_{2,q}(x_1,x_2) & \cdots & \cdots & \ddot{\mathcal{A}}_{m-3,q}(x_1,x_2) & \ddot{\mathcal{A}}_{m,q}(x_1,x_2) \\
\mathcal{B}_{0,q} & \mathcal{B}_{1,q} & \mathcal{B}_{2,q} & \cdots & \cdots & \mathcal{B}_{m-3,q} & \mathcal{B}_{m-1,q} \\
0 & \mathcal{B}_{0,q} & \begin{bmatrix}2\\1\end{bmatrix}_q \mathcal{B}_{1,q} & \cdots & \cdots & \begin{bmatrix}m-3\\1\end{bmatrix}_q \mathcal{B}_{m-4,q} & \begin{bmatrix}m\\1\end{bmatrix}_q \mathcal{B}_{m-2,q} \\
0 & 0 & \mathcal{B}_{0,q} & \cdots & \cdots & \begin{bmatrix}m-3\\2\end{bmatrix}_q \mathcal{B}_{m-5,q} & \begin{bmatrix}m\\2\end{bmatrix}_q \mathcal{B}_{m-2,q} \\
\vdots & \vdots & \vdots & \ddots & \vdots & \vdots & \vdots \\
0 & 0 & 0 & \cdots & \cdots & \mathcal{B}_{0,q} & \begin{bmatrix}m-1\\m-2\end{bmatrix}_q \mathcal{B}_{2,q}
\end{vmatrix}
$$

$$= \frac{-1}{\mathcal{B}_{0,q}} \begin{bmatrix} m \\ m-1 \end{bmatrix}_q \mathcal{B}_{1,q} \mathcal{A}^{[2]}_{m-1,q}(x_1,x_2) - \frac{-1}{(\mathcal{B}_{0,q})} \begin{bmatrix} m-1 \\ m-2 \end{bmatrix}_q \mathcal{B}_{2,q} \mathcal{A}^{[2]}_{m-2,q}(x_1,x_2) + \frac{(-1)^{m-2}}{(\mathcal{B}_{0,q})^{m-1}}$$

$$\begin{vmatrix} 1 & \ddot{\mathcal{A}}_{1,q}(x_1,x_2) & \ddot{\mathcal{A}}_{2,q}(x_1,x_2) & \cdots & \cdots & \ddot{\mathcal{A}}_{m-3,q}(x_1,x_2) & \ddot{\mathcal{A}}_{m,q}(x_1,x_2) \\ \mathcal{B}_{0,q} & \mathcal{B}_{1,q} & \mathcal{B}_{2,q} & \cdots & \cdots & \mathcal{B}_{m-3,q} & \mathcal{B}_{m-1,q} \\ 0 & \mathcal{B}_{0,q} & \begin{bmatrix}2\\1\end{bmatrix}_q \mathcal{B}_{1,q} & \cdots & \cdots & \begin{bmatrix}m-3\\1\end{bmatrix}_q \mathcal{B}_{m-4,q} & \begin{bmatrix}m\\1\end{bmatrix}_q \mathcal{B}_{m-2,q} \\ 0 & 0 & \mathcal{B}_{0,q} & \cdots & \cdots & \begin{bmatrix}m-3\\2\end{bmatrix}_q \mathcal{B}_{m-5,q} & \begin{bmatrix}m\\2\end{bmatrix}_q \mathcal{B}_{m-2,q} \\ \vdots & \vdots & \vdots & \ddots & \vdots & \vdots & \vdots \\ 0 & 0 & 0 & \cdots & \cdots & \mathcal{B}_{0,q} & \begin{bmatrix}m-1\\m-2\end{bmatrix}_q \mathcal{B}_{2,q} \end{vmatrix}.$$

Again, we apply the same technique recursively until we arrive at the following consequence:

$$\mathcal{A}^{[2]}_{m,q}(x_1,x_2) = \frac{-1}{\mathcal{B}_{0,q}} \begin{bmatrix} m \\ m-1 \end{bmatrix}_q \mathcal{B}_{1,q} \mathcal{A}^{[2]}_{m-1,q}(x_1,x_2) - \frac{1}{(\mathcal{B}_{0,q})} \begin{bmatrix} m-1 \\ m-2 \end{bmatrix}_q \mathcal{B}_{2,q} \mathcal{A}^{[2]}_{m-2,q}(x_1,x_2)$$

$$- \cdots - \frac{1}{(\mathcal{B}_{0,q})^2} \begin{vmatrix} 1 & \ddot{\mathcal{A}}_{n,q}(x_1,x_2) \\ \mathcal{B}_{0,q} & \mathcal{B}_{m,q} \end{vmatrix}$$

$$= \frac{-1}{\mathcal{B}_{0,q}} \begin{bmatrix} m \\ m-1 \end{bmatrix}_q \mathcal{B}_{1,q} \mathcal{A}^{[2]}_{m-1,q}(x_1,x_2) - \frac{1}{(\mathcal{B}_{0,q})} \begin{bmatrix} m-1 \\ m-2 \end{bmatrix}_q \mathcal{B}_{2,q} \mathcal{A}^{[2]}_{m-2,q}(x_1,x_2)$$

$$- \cdots - \frac{1}{(\mathcal{B}_{0,q})} \mathcal{B}_{m,q} \mathcal{A}^{[2]}_{0,q}(x_1,x_2) + \frac{1}{\mathcal{B}_{0,q}} \ddot{\mathcal{A}}_{n,q}(x_1,x_2). \tag{68}$$

Finally, upon summing up the series in the left-hand side of the Equation (68), we arrive at the assertion (67) of Theorem 8. $\square$

Corollary 1. *The following identity for the 2DqAP $\ddot{\mathcal{A}}_{n,q}(x_1,x_2)$ holds true:*

$$\ddot{\mathcal{A}}_{m,q}(x_1,x_2) = \sum_{s=0}^{m} \begin{bmatrix} m \\ s \end{bmatrix}_q \mathcal{B}_{m-s,q} \, \mathcal{A}^{[2]}_{k,q}(x_1,x_2) \qquad (m \in \mathbb{N}). \tag{69}$$

4. Several Members of the Twice-Iterated 2D q-Appell Polynomials

During the last two decades, much research work has been done for different members of the family of the q-Appell polynomials and the 2D q-Appell polynomials. By making suitable selections for the functions $\mathcal{A}_q(t)$ and $\ddot{\mathcal{A}}_q(t)$, the members belonging to the family of the twice-iterated 2D q-Appell polynomials $\mathcal{A}^{[2]}_{k,q}(x_1,x_2)$ can be obtained. The 2D q-Bernoulli polynomials $\mathfrak{B}_{m,q}(x_1,x_2)$, the 2D q-Euler polynomials $\mathcal{E}_{m,q}(x_1,x_2)$ and the 2D q-Genocchi polynomials $\mathcal{G}_{m,q}(x_1,x_2)$ are important members of the 2D q-Appell family. Therefore, in this section, we first introduce the 2D q-Euler based Bernoulli polynomials (2DqEBP) $_{\mathcal{E}}\mathfrak{B}_{m,q}(x_1,x_2)$ and the 2D q-Genocchi based Bernoulli polynomials (2DqGBP) $_{\mathcal{G}}\mathfrak{B}_{m,q}(x_1,x_2)$ by means of their respective generating functions and series definitions. We then explore other properties of these members.

4.1. The 2D q-Euler–Bernoulli Polynomials

Since, for

$$A_q(t) = \frac{2}{e_q(t)+1} \quad \text{and} \quad A_q(t) = \frac{t}{e_q(t)-1},$$

the 2DqAP $\mathcal{A}_{m,q}(x_1, x_2)$ reduce to the 2DqEP $\mathcal{E}_{m,q}(x_1, x_2)$ and the 2DqBP $\mathcal{B}_{m,q}(x_1, x_2)$, respectively. Therefore, for the same choices of $A_q(t)$, that is,

$$\dot{A}_q(t) = \frac{2}{e_q(t)+1} \quad \text{and} \quad \ddot{A}_q(t) = \frac{t}{e_q(t)-1},$$

the 2I2DqAP reduce to 2DqEBP $_\mathcal{E}\mathcal{B}_{m,q}(x_1, x_2)$ and are defined by means of generating functions as follows:

$$\frac{2t}{(e_q(t)+1)(e_q(t)-1)}\, e_q(x_1 t) E_q(x_2 t) = \sum_{m=0}^{\infty} {}_\mathcal{E}\mathcal{B}_{m,q}(x_1, x_2)\, \frac{t^m}{[m]_q!} \quad (0 < q < 1). \tag{70}$$

The 2DqEBP $_\mathcal{E}\mathcal{B}_{m,q}(x_1, x_2)$ of degree m are defined by the following series:

$$_\mathcal{E}\mathcal{B}_{m,q}(x_1, x_2) = \sum_{s=0}^{m} \begin{bmatrix} m \\ s \end{bmatrix}_q \mathcal{B}_{s,q}\mathcal{E}_{m-s,q}(x_1, x_2). \tag{71}$$

The following relation between the 2DqEBP $_\mathcal{E}\mathcal{B}_{m,q}(x_1, x_2)$ and the qEBP $_\mathcal{E}\mathcal{B}_{m,q}(x_1)$ holds true:

$$_\mathcal{E}\mathcal{B}_{m,q}(x_1, x_2) = \sum_{s=0}^{m} \begin{bmatrix} m \\ s \end{bmatrix}_q q^{\frac{1}{2}s(s-1)} x_2^s \, {}_\mathcal{E}\mathcal{B}_{m-s,q}(x_1), \tag{72}$$

which, for $x_2 = 1$, yields

$$_\mathcal{E}\mathcal{B}_{m,q}(x_1, 1) = \sum_{s=0}^{m} \begin{bmatrix} m \\ s \end{bmatrix}_q q^{\frac{1}{2}s(s-1)} \, {}_\mathcal{E}\mathcal{B}_{m-s,q}(x_1). \tag{73}$$

The 2DqEBP $_\mathcal{E}\mathcal{B}_{m,q}(x_1, x_2)$ satisfy the following recurrence relation:

$$_\mathcal{E}\mathcal{B}_{m,q}(qx_1, x_2) = q^{m-2}$$
$$\cdot \left(\frac{t(e_q(qt)-1)(x_2(e_q(qt)+1)-e_q(t))+(e_q(t)-te_q(t)-1)(e_q(t)+1)}{t(e_q(t)+1)(e_q(t)-1)} + q^2 x_1 \right) {}_\mathcal{E}\mathcal{B}_{m-1,q}(x_1, x_2). \tag{74}$$

Further, by taking

$$\mathcal{B}_{0,q} = 1,$$

$$\mathcal{B}_{j,q} = \frac{1}{[j+1]} \quad (j \in \mathbb{N})$$

and

$$\ddot{A}_{m,q}(x_1, x_2) = \mathcal{E}_{m,q}(x_1, x_2)$$

in the Equation (56), we obtain the determinant definition of the 2DqEBP $_\mathcal{E}\mathcal{B}_{m,q}(x_1, x_2)$ as given below.

Definition 1. *The 2D q-Euler–Bernoulli polynomials $_{\mathcal{E}}\mathfrak{B}_{m,q}(x_1, x_2)$ of degree m are defined by*

$$_{\mathcal{E}}\mathfrak{B}_{0,q}(x_1, x_2) = 1, \tag{75}$$

$$_{\mathcal{E}}\mathfrak{B}_{m,q}(x_1, x_2) = (-1)^m \begin{vmatrix} 1 & \mathcal{E}_{1,q}(x_1, x_2) & \mathcal{E}_{2,q}(x_1, x_2) & \cdots & \mathcal{E}_{m-1,q}(x_1, x_2) & \mathcal{E}_m(x_1, x_2) \\ 1 & \frac{1}{[2]_q} & \frac{1}{[3]_q} & \cdots & \frac{1}{[m]_q} & \frac{1}{[m+1]_q} \\ 0 & 1 & \begin{bmatrix} 2 \\ 1 \end{bmatrix}_q \frac{1}{[2]_q} & \cdots & \begin{bmatrix} m-1 \\ 1 \end{bmatrix}_q \frac{1}{[m-1]_q} & \begin{bmatrix} m \\ 1 \end{bmatrix}_q \frac{1}{[m]_q} \\ 0 & 0 & 1 & \cdots & \begin{bmatrix} m-1 \\ 2 \end{bmatrix}_q \frac{1}{[m-2]_q} & \begin{bmatrix} m \\ 2 \end{bmatrix}_q \frac{1}{[m-1]_q} \\ \vdots & \vdots & \vdots & \ddots & \vdots & \vdots \\ 0 & 0 & 0 & \cdots & 1 & \begin{bmatrix} m \\ m-1 \end{bmatrix}_q \frac{1}{[2]_q} \end{vmatrix} \tag{76}$$

$$(m \in \mathbb{N}),$$

where $\mathcal{E}_{m,q}(x_1, x_2)$ $(m \in \mathbb{N}_0)$ are the 2D q-Euler polynomials of degree m.

4.2. The 2D q-Genocchi–Bernoulli Polynomials

Since, for

$$\mathcal{A}_q(t) = \frac{2t}{e_q(t) + 1} \qquad \text{and} \qquad \mathcal{A}_q(t) = \frac{t}{e_q(t) - 1},$$

the 2DqAP $\mathcal{A}_{m,q}(x_1, x_2)$ reduce to the 2DqGP $\mathcal{G}_{m,q}(x_1, x_2)$ and the 2DqBP $\mathfrak{B}_{m,q}(x_1, x_2)$, respectively. Therefore, for the same choices of $A_q(t)$, that is,

$$\grave{A}_q(t) = \frac{2t}{e_q(t) + 1} \qquad \text{and} \qquad \ddot{A}_q(t) = \frac{t}{e_q(t) - 1},$$

the 2I2DqAP reduce to 2DqGBP $_{\mathcal{G}}\mathfrak{B}_{m,q}(x_1, x_2)$ and are defined by means of generating functions as follows:

$$\frac{2t^2}{(e_q(t) + 1)(e_q(t) - 1)} e_q(x_1 t) E_q(x_2 t) = \sum_{m=0}^{\infty} {}_{\mathcal{G}}\mathfrak{B}_{m,q}(x_1, x_2) \frac{t^m}{[m]_q!} \qquad (0 < q < 1). \tag{77}$$

The 2DqGBP $_{\mathcal{G}}\mathfrak{B}_{m,q}(x_1, x_2)$ of degree m are defined by the following series:

$$_{\mathcal{G}}\mathfrak{B}_{m,q}(x_1, x_2) = \sum_{s=0}^{m} \begin{bmatrix} m \\ s \end{bmatrix}_q \mathfrak{B}_{s,q} \, \mathcal{G}_{m-s,q}(x_1, x_2). \tag{78}$$

The following relation between the 2DqGBP $_{\mathcal{G}}\mathfrak{B}_{m,q}(x_1, x_2)$ and the qGBP $_{\mathcal{G}}\mathfrak{B}_{m,q}(x_1)$ holds true:

$$_{\mathcal{G}}\mathfrak{B}_{m,q}(x_1, x_2) = \sum_{s=0}^{m} \begin{bmatrix} m \\ s \end{bmatrix}_q q^{\frac{1}{2}s(s-1)} x_2^s \, {}_{\mathcal{G}}\mathfrak{B}_{m-s,q}(x_1), \tag{79}$$

which, for $x_2 = 1$, gives

$$_{\mathcal{G}}\mathfrak{B}_{m,q}(x_1, 1) = \sum_{s=0}^{m} \begin{bmatrix} m \\ s \end{bmatrix}_q q^{\frac{1}{2}s(s-1)} \, {}_{\mathcal{G}}\mathfrak{B}_{m-s,q}(x_1). \tag{80}$$

The 2DqGBP $_G\mathfrak{B}_{m,q}(x_1, x_2)$ satisfy the following recurrence relation:

$$
_G\mathfrak{B}_{m,q}(qx_1, x_2) = q^{m-3} \cdot \left(\frac{(e_q(qt)+1)(e_q(t) - te_q(t) + 1 + x_2 t(e_q(qt)+1))}{t(e_q(t)+1)(e_q(t)-1)} \right.
$$
$$
\left. + \frac{q(e_q(t) - te_q(t) - 1)(e_q(t)+1)}{t(e_q(t)+1)(e_q(t)-1)} + q^3 x_1 \right) {}_G\mathfrak{B}_{m-1,q}(x_1, x_2). \quad (81)
$$

In the next section (Section 5 below), we give some graphical representations and the surface plots of some of the members of the twice-iterated 2D q-Appell polynomials.

5. Graphical Representations and Surface Plots

Here, in this section, the graphs of the q-Euler–Bernoulli polynomials (qEBP) $_\mathcal{E}\mathfrak{B}_{m,q}(x)$, q-Genocchi-Bernoulli polynomials (qGBP) $_G\mathfrak{B}_{m,q}(x)$ and the surface plots of the 2DqEBP $_\mathcal{E}\mathfrak{B}_{m,q}(x_1, x_2)$ and the 2DqGBP $_G\mathfrak{B}_{m,q}(x_1, x_2)$ are presented.

To draw the plot of the qEBP $_\mathcal{E}\mathfrak{B}_{m,q}(x)$ and the qGBP $_G\mathfrak{B}_{m,q}(x)$, we choose $q = \frac{1}{2}$ and consider the values of the first four q-Euler–Bernoulli polynomials and of the first four q-Genocchi–Bernoulli polynomials, the expressions of these polynomials are given in Table 1.

Table 1. Expressions of the first four $_\mathcal{E}\mathfrak{B}_{m,\frac{1}{2}}(x)$ and $_G\mathfrak{B}_{m,\frac{1}{2}}(x)$.

m	0	1	2	3	3
$_\mathcal{E}\mathfrak{B}_{m,\frac{1}{2}}(x)$	1	$x - \frac{7}{6}$	$x^2 - \frac{7}{4}x + \frac{79}{168}$	$x^3 - \frac{49}{24}x^2 + \frac{79}{96}x + \frac{379}{2880}$	$x^4 - \frac{35}{16}x^3 + \frac{145}{192}x^2 + \frac{379}{1536}x + .0213$
$_G\mathfrak{B}_{m,\frac{1}{2}}(x)$	0	1	$\frac{3}{2}x - \frac{7}{4}$	$\frac{7}{4}x^2 - \frac{49}{16}x + \frac{121}{96}$	$\frac{15}{8}x^3 - \frac{45}{16}x^2 + \frac{815}{256}x + \frac{379}{1536}$

Further, by setting $m = 4$ and $q = \frac{1}{2}$ in the series definitions (72) and (79) of $_\mathcal{E}\mathfrak{B}_{m,\frac{1}{2}}(x_1, x_2)$ and $_G\mathfrak{B}_{m,q}(x_1, x_2)$ and using the particular values of $_\mathcal{E}\mathfrak{B}_{m,\frac{1}{2}}(x)$ and $_G\mathfrak{B}_{m,\frac{1}{2}}(x)$ from Table 1, we find that

$$
\begin{aligned}
_\mathcal{E}\mathfrak{B}_{4,\frac{1}{2}}(x_1, x_2) =& x_1^4 - \frac{35}{16}x_1^3 + \frac{145}{192}x_1^2 + \frac{379}{1536}x_1 + 0.0213 + \frac{15}{8}x_1^3 x_2 - \frac{245}{64}x_1^2 x_2 + \frac{395}{256}x_1 x_2 \\
& + \frac{379}{1536}x_2 + \frac{35}{32}x_1^2 x_2^2 - \frac{245}{128}x_1 x_2^2 + \frac{395}{768}x_2^2 + \frac{15}{64}x_1 x_2^3 - \frac{35}{128}x_2^3 + \frac{1}{64}x_2^4
\end{aligned} \quad (82)
$$

and

$$
\begin{aligned}
_G\mathfrak{B}_{3,\frac{1}{2}}(x_1, x_2) =& \frac{15}{8}x^3 - \frac{45}{16}x^2 + \frac{815}{256}x + \frac{379}{1536} + \frac{105}{32}x_1^2 x_2 - \frac{735}{128}x_1 x_2 - \frac{605}{256}x_2 \\
& + \frac{105}{64}x_1 x_2^2 - \frac{245}{128}x_2^2 + \frac{15}{64}x_2^3.
\end{aligned} \quad (83)
$$

Next, by using the expression given in Table 1 and the Equations (82) and (83), with the help of Matlab, we get the Figures 1–4 below.

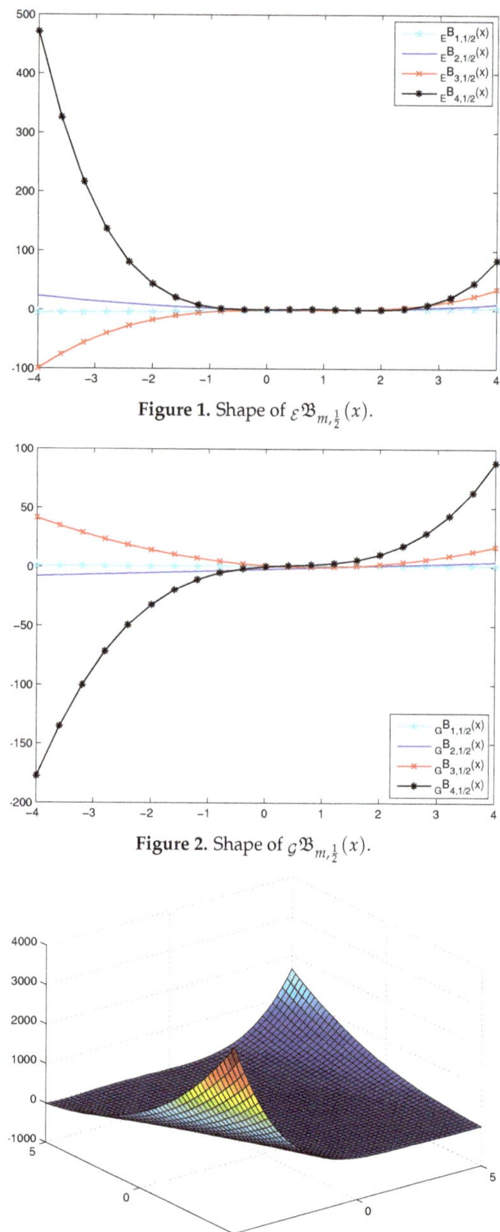

Figure 1. Shape of $\mathcal{E}\mathfrak{B}_{m,\frac{1}{2}}(x)$.

Figure 2. Shape of $\mathcal{G}\mathfrak{B}_{m,\frac{1}{2}}(x)$.

Figure 3. Surface plot of $\mathcal{E}\mathfrak{B}_{4,\frac{1}{2}}(x_1, x_2)$.

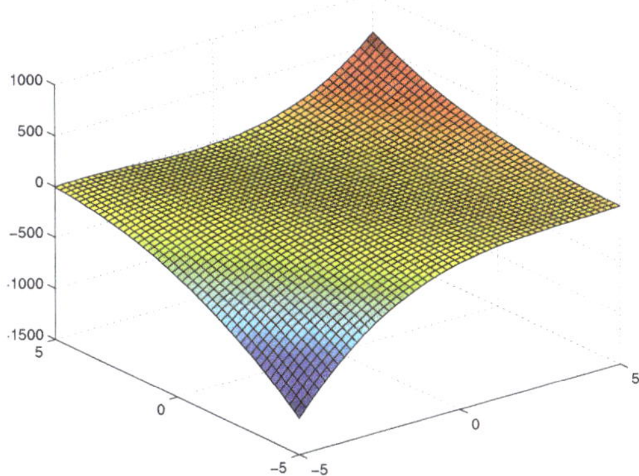

Figure 4. Surface plot of $_g\mathfrak{B}_{4,\frac{1}{2}}(x_1, x_2)$.

Further, with the help of Matlab, we compute the real and complex zeros of $_\varepsilon\mathfrak{B}_{m,\frac{1}{2}}(x)$ and $_g\mathfrak{B}_{m,\frac{1}{2}}(x)$ for $m = 1, 2, 3, 4$ and $x \in \mathbb{C}$. These zeros are mentioned in Tables 2 and 3.

Table 2. Real zeros of $_\varepsilon\mathfrak{B}_{m,\frac{1}{2}}(x)$ and $_g\mathfrak{B}_{m,\frac{1}{2}}(x)$.

m	$_\varepsilon\mathfrak{B}_{m,\frac{1}{2}}(x)$	$_g\mathfrak{B}_{m,\frac{1}{2}}(x)$
1	1.1667	0
2	0.3315, 1.4185	1.1667
3	$-0.1213, 0.7910, 1.3719$	0.6620, 1.0880
4	0.7878, 1.6239	-0.0726

Table 3. Complex zeros of $_\varepsilon\mathfrak{B}_{m,\frac{1}{2}}(x)$ and $_g\mathfrak{B}_{m,\frac{1}{2}}(x)$.

m	$_\varepsilon\mathfrak{B}_{m,\frac{1}{2}}(x)$	$_g\mathfrak{B}_{m,\frac{1}{2}}(x)$
1	—	—
2	—	—
3	—	—
4	$-0.1121 - 0.0639i, -0.1121 + 0.0639i$	$0.7863 - 1.0926i, 0.7863 + 1.0926i$

Also, with the help of Matlab, the zeros mentioned in Tables 2 and 3 are shown in the Figures 5 and 6.

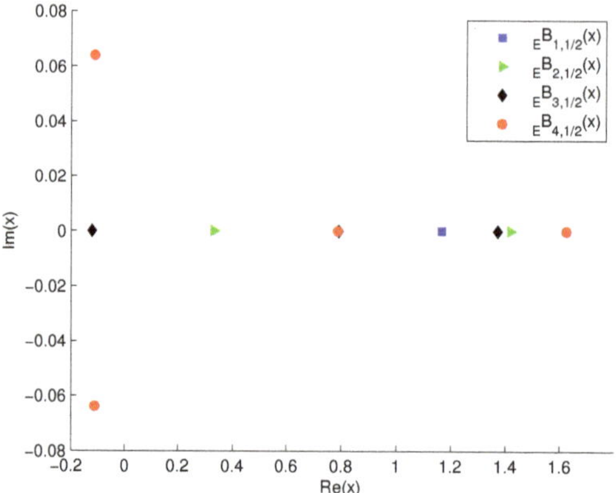

Figure 5. Zeros of $_{\mathcal{E}}\mathfrak{B}_{m,\frac{1}{2}}(x)$.

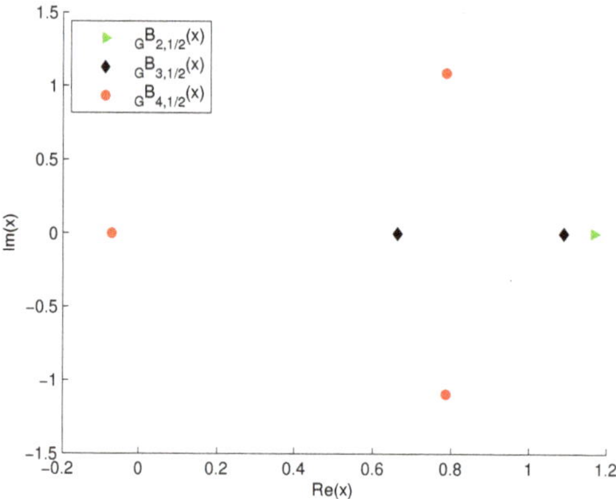

Figure 6. Zeros of $_{\mathcal{G}}\mathfrak{B}_{m,\frac{1}{2}}(x)$.

6. Concluding Remarks and Observations

As long ago as 1910, Jackson [27] studied the q-definite integral of an arbitrary function $f(t)$, which is defined as follows:

$$\int_0^a f(t)\, d_q t = (1-q)a \sum_{m=0}^{\infty} q^m f(aq^m) \qquad (0 < q < 1;\ a \in \mathbb{R}) \tag{84}$$

and

$$\int_a^b f(t) d_q t = \int_0^b f(t) d_q t - \int_0^a f(t)\, d_q t. \tag{85}$$

We note also that

$$D_q \int_0^t f(x) d_q x = f(t). \tag{86}$$

Applying the double q-integral to both sides of the Equation (42), that is,

$$\int_0^{x_1} \int_0^{x_2} [m]_q \, \mathcal{A}^{[2]}_{m-1,q}(t_1, qt_2) \, d_q t_1 \, d_q t_2 = \int_0^{x_1} \int_0^{x_2} D_{q,t_2} \mathcal{A}^{[2]}_{m,q}(t_1, t_2) \, d_q t_1 \, d_q t_2, \tag{87}$$

we have

$$[m]_q \int_0^{x_1} \int_0^{x_2} \mathcal{A}^{[2]}_{m-1,q}(t_1, qt_2) \, d_q t_1 \, d_q t_2 = \int_0^{x_1} \left(\mathcal{A}^{[2]}_{m,q}(t_1, x_2) - \mathcal{A}^{[2]}_{m,q}(t_1, 0) \right) d_q t_1. \tag{88}$$

In view of the Equation (41), the above Equation (88) yields

$$[m]_q \int_0^{x_1} \int_0^{x_2} \mathcal{A}^{[2]}_{m-1,q}(t_1, qt_2) \, d_q t_1 d_q t_2$$
$$= \int_0^{x_1} \frac{1}{[m+1]_q} \left(D_{q,t_1} \mathcal{A}^{[2]}_{m+1,q}(t_1, x_2) - D_{q,t_1} \mathcal{A}^{[2]}_{m+1,q}(t_1, 0) \right) d_q t_1 \tag{89}$$
$$= \frac{1}{[m+1]_q} \left(\mathcal{A}^{[2]}_{m+1,q}(x_1, x_2) - \mathcal{A}^{[2]}_{m+1,q}(0, x_2) - \mathcal{A}^{[2]}_{m+1,q}(x_1, 0) + \mathcal{A}^{[2]}_{m+1,q}(0, 0) \right),$$

which, on using the Equations (13) and (39), becomes

$$\int_0^{x_1} \int_0^{x_2} \mathcal{A}^{[2]}_{m,q}(t_1, qt_2) \, d_q t_1 d_q t_2$$
$$= \frac{1}{[m+1]_q [m+2]_q} \left(\mathcal{A}^{[2]}_{m+2,q}(x_1, x_2) - (-1)^m \mathcal{A}^{[2]}_{m+2,q}(x_1, -x_2) - \mathcal{A}^{[2]}_{m+2,q}(x_1) + \mathcal{A}^{[2]}_{m+2,q} \right). \tag{90}$$

Further, in view of the Equations (31) and (34), the Equations (90) yields

$$\int_0^{x_1} \int_0^{x_2} \mathcal{A}^{[2]}_{m,q}(t_1, qt_2) \, d_q t_1 d_q t_2 = \frac{1}{[m+1]_q [m+2]_q}$$
$$\cdot \sum_{s=0}^{m+2} \begin{bmatrix} m+2 \\ s \end{bmatrix}_q \mathcal{A}_{s,q} \left(\ddot{\mathcal{A}}_{m+2-s,q}(x_1, x_2) - (-1)^m \ddot{\mathcal{A}}_{m+2-s,q}(x_1, x_2) - \ddot{\mathcal{A}}_{m+2-s,q}(x_1) + \ddot{\mathcal{A}}_{m+2-s,q} \right). \tag{91}$$

In conclusion, we choose to reiterate the now well-understood fact that the results for the q-analogues, which we have considered in this article for $0 < q < 1$, can easily be translated into the corresponding results for the so-called (p, q)-analogues (with $0 < q < p \leqq 1$) by applying some obviously trivial parametric and argument variations, the additional parameter p being redundant. In fact, the so-called (p, q)-number $[n]_{p,q}$ is given (for $0 < q < p \leqq 1$) by (see also [28])

$$[n]_{p,q} := \begin{cases} \dfrac{p^n - q^n}{p - q} & (n \in \{1, 2, 3, \cdots\}) \\[2ex] 0 & (n = 0) \end{cases}$$
$$=: p^{n-1} \, [n]_{\frac{q}{p}}, \tag{92}$$

where, for the classical q-number $[n]_q$, we have

$$[n]_q := \frac{1 - q^n}{1 - q}$$
$$= p^{1-n} \left(\frac{p^n - (pq)^n}{p - (pq)} \right) \tag{93}$$
$$= p^{1-n} \, [n]_{p, pq}.$$

Consequently, any claimed extensions of most (including the present) investigations involving the classical q-calculus to the corresponding obviously straightforward investigations involving the (p, q)-calculus are truly inconsequential.

Further investigations along the lines presented in this paper, which are associated with the various recent generalizations and extensions of the Apostol type Bernoulli, Euler and Genocchi polynomials introduced by, for example, Srivastava et al. (see [29,30]) may be worthy of consideration by the targeted readers.

Author Contributions: All authors contributed equally.

Funding: This research received no external funding.

Conflicts of Interest: The authors declare no conflict of interest.

References

1. Strominger, A. Information in black hole radiation. *Phys. Rev. Lett.* **1993**, *71*, 3743–3746.
2. Youm, D. q-deformed conformal quantum mechanics. *Phys. Rev. D* **2000**, *62*, 095009. [CrossRef]
3. Lavagno, A.; Swamy, P.N. q-deformed structures and nonextensive statistics: A comparative study. *Phys. A Stat. Mech. Appl.* **2002**, *305*, 310–315. [CrossRef]
4. Andrews, G.E.; Askey, R.; Roy, R. *71st Special Functions of Encyclopedia of Mathematics and Its Applications*; Cambridge University Press: Cambridge, UK, 1999.
5. Aral, A.; Gupta, V.; Agarwal, R.P. *Applications of q-Calculus in Operator Theory*; Springer: New York, NY, USA, 2013.
6. Ernst, T. *A Comprehensive Treatment of q-Calculus*; Springer: Basel, Switzerland; Heidelberg, Germany; New York, NY, USA; Dordrecht, The Netherlands; London, UK, 2012.
7. Appell, P. Sur une classe de polynômes. *Ann. Sci. École. Norm. Supér.* **1880**, *9*, 119–144. [CrossRef]
8. Thorne, C.J. A property of Appell sets. *Am. Math. Mon.* **1945**, *52*, 191–193. [CrossRef]
9. Sheffer, I.M. Note on Appell polynomials. *Bull. Am. Math. Soc.* **1945**, *51*, 739–744. [CrossRef]
10. Varma, R.S. On Appell polynomials. *Proc. Am. Math. Soc.* **1951**, *2*, 593–596. [CrossRef]
11. Pintér, Á.; Srivastava, H.M. Addition theorems for the Appell polynomials and the associated classes of polynomial expansions. *Aequ. Math.* **2013**, *85*, 483–495. [CrossRef]
12. Srivastava, H.M.; Özarslan, M.A.; Yılmaz, B. Some families of differential equations associated with the Hermite-based Appell polynomials and other classes of Hermite-based polynomials. *Filomat* **2014**, *28*, 695–708. [CrossRef]
13. Srivastava, H.M.; Özarslan, M.A.; Yaşar, B.Y. Difference equations for a class of twice-iterated Δ_h-Appell sequences of polynomials. *Rev. Real Acad. Cienc. Exactas Fís. Natur. Ser. A Mat. (RACSAM)* **2019**, *113*, 1851–1871. [CrossRef]
14. Srivastava, H.M.; Ricci, P.E.; Natalini, P. A family of complex Appell polynomial sets. *Rev. Real Acad. Cienc. Exactas Fís. Nat. Ser. A Mater. (RACSAM)* **2019**, *113*, 2359–2371. [CrossRef]
15. Sharma, A.; Chak, A.M. The basic analogue of a class of polynomials. *Riv. Mat. Univ. Parma (Ser. 4)* **1954**, *5*, 325–337.
16. Al-Salam, W.A. q-Appell polynomials. *Ann. Mat. Pura Appl. (Ser. 4)* **1967**, *4*, 31–45. [CrossRef]
17. Srivastava, H.M. Some characterizations of Appell and q-Appell polynomials. *Ann. Mat. Pura Appl. (Ser. 4)* **1982**, *130*, 321–329. [CrossRef]
18. Anshelevich, M. Appell polynomials and their relatives. III: Conditionally free theory. *Ill. J. Math.* **2009**, *53*, 39–66. [CrossRef]
19. Levi, D.; Tempesta, P.; Winternitz, P. Umbral calculus, difference equations and the discrete Schrödinger equation. *Math. Phys.* **2004**, *45*, 4077–4105. [CrossRef]
20. Tempesta, P. Formal groups, Bernoulli-type polynomials and L-series. *C. R. Math. Acad. Sci. Paris* **2007**, *345*, 303–306. [CrossRef]

21. Keleshteri, M.E.; Mahmudov, N.I. A study on q-Appell polynomials from determinant point of view. *Appl. Math. Comput.* **2015**, *260*, 351–369.

22. Srivastava, H.M.; Khan, S.; Riyasat, M. q-Difference equations for the twice-iterated q-Appell and mixed type q-Appell Polynomials. *Arab. J. Math.* **2019**, *8*, 63–77. [CrossRef]

23. Carlitz, L. q-Bernoulli numbers and polynomials. *Duke Math. J.* **1948**, *15*, 987–1000. [CrossRef]

24. Mahmudov, N.I. On a class of q-Bernoulli and q-Euler polynomials. In: Proceedings of the International Congress in Honour of Professor Hari M. Srivastava. *Adv. Differ. Equ.* **2013**, *2013*, 108. [CrossRef]

25. Srivastava, H.M. Some generalizations and basic (or q-) extensions of the Bernoulli, Euler and Genocchi polynomials. *Appl. Math. Inf. Sci.* **2011**, *5*, 390–444.

26. Khan, S.; Riyasat, M. Determinant approach to the twice-iterated q-Appell and mixed type q-Appell polynomials. *arXiv* **2016**, arXiv:1606.04265.

27. Jacson, F.H. On q-definite integrals. *Q. J. Pure Appl. Math.* **1910**, *41*, 193–203.

28. Srivastava, H.M.; Tuglu, N.; Çetin, M. Some results on the q-analogues of the incomplete Fibonacci and Lucas polynomials. *Miskolc Math. Notes* **2019**, *20*, 511–524. [CrossRef]

29. Srivastava, H.M.; Masjed-Jamei, M.; Beyki, M.R. A parametric type of the Apostol-Bernoulli, Apostol-Euler and Apostol-Genocchi polynomials. *Appl. Math. Inf. Sci.* **2018**, *12*, 907–916. [CrossRef]

30. Srivastava, H.M.; Masjed-Jamei, M.; Beyki, M.R. Some new generalizations and applications of the Apostol-Bernoulli, Apostol-Euler and Apostol-Genocchi polynomials. *Rocky Mt. J. Math.* **2019**, *49*, 681–697. [CrossRef]

© 2019 by the authors. Licensee MDPI, Basel, Switzerland. This article is an open access article distributed under the terms and conditions of the Creative Commons Attribution (CC BY) license (http://creativecommons.org/licenses/by/4.0/).

MDPI

St. Alban-Anlage 66

4052 Basel

Switzerland

Tel. +41 61 683 77 34

Fax +41 61 302 89 18

www.mdpi.com

Symmetry Editorial Office

E-mail: symmetry@mdpi.com

www.mdpi.com/journal/symmetry

www.ingramcontent.com/pod-product-compliance
Lightning Source LLC
LaVergne TN
LVHW070703170726
843515LV00004B/472